ECOLOGIES OF COMPARISON

EXPERIMENTAL FUTURES
Technological Lives, Scientific Arts, Anthropological Voices
A series edited by Michael M. J. Fischer and Joseph Dumit

ECOLOGIES OF COMPARISON

An Ethnography of Endangerment in Hong Kong

TIM CHOY

 UNIVERSITY PRESS

DURHAM AND LONDON   2011

CONTENTS

ACKNOWLEDGMENTS

This book and I have grown together for some time. We share an itinerary through Santa Cruz, Hong Kong, Columbus, Ithaca, Davis, and the San Francisco Bay Area, and we owe our form to many people, collectives, and institutions in these and other places.

Wonderful mentors advised me in graduate school as I developed this project. Anna Tsing encouraged slow thinking and pressed assumptions of scale. Donna Haraway helped me glimpse how ecology, ethnography, and theory could shape one another. Lisa Rofel reminded me of the critical functions of nonfiction and argument. David Hoy taught me to see philosophies as toolkits. Hugh Raffles reminded me to write. I hope they will recognize my gratitude to them throughout this work.

This book could not have been written without the generosity of those who welcomed me into their professional and personal lives in Hong Kong. I am especially grateful to the staff of Greenpeace, Hong Kong, particularly Chan Yiu Kwong, Rupert Yu, Fan Wing Hung, and Maria Deng, for allowing me to observe and participate in their actions and meetings. I thank the many people in Tai O, Lung Kwu Tan, and Ha Pak Nai who spoke frankly with me about their predicaments and about politics. I owe special thanks to Wong Wai King, as well as to Kuo Chun Chuen, Rosanna, and Jenny at the Tai O YWCA Community Center, for facilitating my research in Tai O. Though I must keep their names and affiliations confidential, I also thank the environmental engineers and consultants with whom I worked, as well as the officials in the Hong Kong Environmental Protection Department and Planning Bureau; all impressed me with their candor and generosity. I thank Mei Sun for encouraging me to work with her story and for a last-minute correction.

While conducting field research, I enjoyed the privilege of affiliation with the Centre of Asian Studies at the University of Hong Kong. Elizabeth Sinn was a keen and insightful host, offering timely advice, helpful introductions, and the occasional pep talk. Pun Ngai was a valued interlocutor in the Hong Kong Culture and Society Program, as were HKCSP staff members Aris Chan, Yinha Chan, Ah Chong, and Pik-ching Ip. I am also grateful to Man Si-wai and Stephen Chiu of the Chinese University of Hong Kong, Cheung Siu-woo of the Hong Kong University of Science and Technology, and John Erni of Hong Kong City University for their encouragement and advice. Edward Stokes showed me a landscape I had not known. Yun-wing Sung provided me with some crucial introductions. Friendships with Clement and Phoebe Lam, Kaming Wu, Lo Sze Ping, Sze Pang-Cheung, Kendy Yim, Lily Lau, Yan Sham-Shackleton, Luisa Tam, Myfanwy Hughes, Tom Boasberg, and Carin Chow made Hong Kong not only a research site but a home. Others who made Hong Kong home were my extended family, the Sungs. To Manling Sung and Wat Kwong San, Yun-wing and Pauline Sung, Lance Sung, and Francis and Jodi Sung, I owe special thanks. They shared stories, included me in family events, and provided a logistical and emotional safety net.

For material support, I am grateful to the National Science Foundation for a Graduate Research Fellowship, to the Hong Kong Culture and Society Program at the University of Hong Kong, and the University of California Pacific Rim Research Program for fieldwork funding. I also thank the UC Office of the President for a President's Graduate Fellowship and Dissertation-Year Fellowship.

I completed a draft of this book while teaching in the Department of Comparative Studies at Ohio State University, and I am grateful for the emboldening environment of critique and collaboration that my colleagues fostered there. Special thanks to Maurice Stevens and Ruby Tapia for writing and reading with me; to Eugene Holland, Daniel Reff, and Barry Shank for their careful reading of chapters in progress; to Phillip Armstrong and Brian Rotman for posing questions no anthropologist would ask; and to Alana Kumbier, Rohit Negi, and Taylor Nelms for reading an early draft in seminar. I am indebted to David Horn, who offered incisive criticism and advice while helping me carve out and protect time for research. I am grateful to Nina Berman, Luz Calvo, Tanya Erzen, Nancy Jesser, Lindsay Jones, Thomas

Kasulis, Kwaku Korang, Rick Livingston, Dorothy Noyes, Hugh Urban, Julia Watson, and Sabra Webber for their intellectual friendship and for making interdisciplinary collegiality ordinary rather than exceptional. Thanks also go to Sylvia McDorman and Lori Wilson for helping me get things done so efficiently that I barely recognized myself.

A fellowship at the Clarke Center for East Asian Culture and Law at Cornell in 2006–7 gave me an opportunity to tune my arguments with fresh audiences and readers. Annelise Riles facilitated my entry into a wonderful year of conversations, readings, and critiques. I am especially grateful to Annelise, Tony Crook, Eduardo Kohn, Stacey Langwick, Rachel Prentice, Hiro Miyazaki, Adam Reed, and Audra Simpson for reading full drafts of the manuscript. I also thank Dominic Boyer, Doug Kysar, Michael Lynch, Eva Pils, Trevor Pinch, and Marina Welker for making me feel welcome in the law, anthropology, and science and technology studies communities at Cornell.

Along the way, Nicholas Brown, Lieba Faier, Michael Fischer, Kim Fortun, Michael Fortun, Cori Hayden, Jeremy Hermann, Heidi Hess, Jake Kosek, Ralph Litzinger, Celia Lowe, Bill Maurer, Michael Montoya, Kris Peterson, Shiho Satsuka, Kaushik Sunder Rajan, and Mei Zhan have sustained me and this book with critique, encouragement, and friendship. I am especially thankful to Stacey Langwick for ongoing inspiration and camaraderie. Thank you all. I am grateful to three anonymous reviewers, whose thoughtful suggestions helped bring out the themes that mattered most, as well as to Theresa MacPhail, who offered a very productive reading on short notice. Throughout the process Ken Wissoker has been a wonderful editor, shepherding me with grace and patience. My sincere thanks also go to Tim Elfenbein and Erin Holloway, my managing editor and copy editor, for tightening my prose and helping me wrap things up.

I finish this book in ideal company at the University of California, Davis. I am deeply grateful to Joe Dumit, who always seems to have read everything already, for his comradeship, wise advice, and good humor. I thank Tom Beamish, Mario Biagioli, Joan Cadden, Patrick Carroll, Marisol de la Cadena, Carolyn de la Peña, Donald Donham, Cristiana Giordano, Jim Griesemer, Suad Joseph, Caren Kaplan, Alan Klima, Colin Milburn, Roberta Millstein, Ben Orlove, Suzana Sawyer, Janet Shibamoto-Smith, James Smith, Smriti Srinivas, Daniel Stolzenberg, Aram Yengoyan, and Li Zhang for fostering atmospheres of experiment and generosity in my two academic homes, Science and Technology Studies and Sociocultural Anthropology. Kelly Byrns, Lesley Byrns, Mary Dixon, Dewight Kramer, Nicole Kramer, Carol McMasters-Stone, Nancy McLaughlin, Edie Stasulat, and Heidi Williams help me navigate our campus, literally and figuratively. Vivian Choi, Nicholas d'Avella, William DiFede, Stefanie Graeter, Chris Kortright, Jieun Lee, Rima Praspaliauskiene, and Michelle Stewart inspire and teach me.

Profound and heartfelt thanks go to my parents, Ben and Louisa Choy. They supported me throughout this project, and I could not have done it without their care and forbearance. Finally, I thank Zamira Ha—my partner every step of the way—for traveling and wondering with me, for incisive critiques, and for showing me how to erase.

• • •

An earlier version of chapter 4 appeared as "Articulated Knowledges: Environmental Forms after Universality's Demise," *American Anthropologist* 107, no. 1 (2005), 5–18.

NOTE ON TRANSLITERATION

Cantonese romanizations in this book generally follow the Yale system but omit non-alphabetical tone diacritics. Exceptions include: street and place names, such as "Hong Kong," which follow government spellings; personal names, which follow people's chosen spellings; and excerpts from published works, which reflect authors' original spellings. Mandarin romanizations are in Pinyin.

PROBLEMS OF A POLITICAL NATURE

One day in 1995, some indigenous clansmen mounted bulldozers and cleared a six-hectare tract of land in Sha Lo Tung valley, the heart of one of Hong Kong's country parks. The bulldozing captivated onlookers and reporters, who assumed the land was protected after a coalition of environmental NGOs successfully stalled a village-backed proposal to build a low-density housing complex there. Furious at the plan's blocking, with neither approval nor compensation for lost development potential in sight, the clansmen took matters into their own hands. As they flattened trees, uprooted vegetation, and tore through soil, they told observers that they were simply preparing the plot of land for agriculture. If they could not build, they would farm—as their ancestors had done. Environmentalists, meanwhile, decried the event as an attempt to destroy the ecological value of the area, a brazen plot to subvert the region's potential scientific interest and therefore its qualification for environmental protection.

I learned of this event a few years after the fact. I remember the moment vividly. I was conducting field research on the global circulations of environmental expertise in Hong Kong, interviewing Janet, a young British expatriate in Hong Kong who had worked for several years as the spokesperson for a prominent international environmental NGO.[1] When I asked her how local people received her work, Janet nodded quickly, remarking that doing environmental politics in Hong Kong required confronting the perception of environmentalism as a foreign or "Western" political platform. It presented a real problem, she admitted, and environmental organizations, particularly international ones, needed to deal with it more effectively. For instance, had I heard of Sha Lo Tung? The most heart-wrenching thing she had witnessed in her career had taken place there.

Janet had been active in the coalition that mobilized to halt development near Sha Lo Tung. Uniting professional environmental NGOs like Friends of the Earth and the Worldwide Fund for Nature, as well as volunteer organizations like Green Power and Green Lantau Association, the coalition emerged in 1990, when the Hong Kong Lands Department approved the Sha Lo Tung development proposal.

When initially submitted to the Hong Kong government in 1979, the proposal included plans for a nine-hole golf course with a nearby country club and residential developments. Over the next ten years, however, a remarkable thing happened. As the proposal was shuffled between the Sha Lo Tung Development Company and various government departments, it grew significantly in scope. By 1990 the proposed project had evolved into a large campus encompassing an eighteen-hole golf course, a country club, sixty-six low-density houses, and two hundred apartments, all encroaching on protected lands in Pat Sin Leng Country Park, one of Hong Kong's many country parks, by 31 hectares.

Developers are usually not allowed to build in a country park, but the Sha Lo Tung Development Company found a loophole: indigenous partners. Under Hong Kong law, men descended through the male line from residents of villages that were recognized in 1898 by the colonial government hold certain land rights, including inheritable ownership and building rights. The Sha Lo Tung Development Company had sought and gained village partners, buying land from villagers in return for a promise of a cut of the profits. While the proposed housing complex would lie within the borders of the country park, technically, it would not be built on park land. In this way, the company could locate luxury housing in the middle of Pat Sin Leng.

Janet and others in the coalition responded instantly to the Hong Kong government's approval of the proposal, coordinating and publicizing a week of petitions, marches, and lobbying efforts, and their actions bore remarkable fruit. Hong Kong's Agriculture and Fisheries Department, which had initially authorized the development, now backpedaled and admitted that its approval had hinged on a technical error. Officials from the Environmental Protection Department remembered that they still had not received an independent environmental impact assessment requested years ago, and researchers began to evaluate the Sha Lo Tung Valley as a potential Site of

Special Scientific Interest. Scientists determined that 65 percent of Hong Kong's dragonfly species inhabited the land around Sha Lo Tung, at least two varieties of which were unique to Hong Kong. If the valley were classified a Site of Special Scientific Interest by the Town Planning Board, it would be off-limits to development. By the mid-1990s environmentalists appeared to have won the day: the golf course and housing complex were stalled. But whatever satisfaction Janet enjoyed from this victory disappeared when Sha Lo Tung villagers mounted their bulldozers and purposely flattened some of the lands—now "habitats"—that she and her colleagues had worked so hard to protect.

I listened raptly to Janet's account, but at her mention of bulldozers, I suddenly realized that I had heard the story before. Shortly after my arrival in Hong Kong, relatives who thought I might appreciate a tale of cultural conflict in environmental controversy had told me about the incident. In fact, I would encounter it several times during my fifteen months in Hong Kong, leading me to wonder how Sha Lo Tung had come to be held so widely as an exemplary case.

Certainly, the bulldozing presented a powerful story. It featured a cast of stereotypical players: environmentalists protecting an undeveloped landscape and some obscure animals from a golf course; villagers decrying the meddling of outsiders; scientists generating new data for use in environmentalist arguments. These players met in a political drama that appeared, as it unfolded, to move toward one outcome but then turned dramatically to another. This plot twist was striking, but what most arrested me came afterward. "And they bulldozed the trees to the ground!" Then nothing.

Silence followed the punch line—an empty beat rousing me from my recollections in Janet's office, a punctuating pause in conversations with family and friends. My interlocutors' eyes would take that beat to scan my face, waiting.

At stake in the silence were two related things. The first was my location and stance: Would I view the example from within or outside environmentalism, from within or outside a commitment to Hong Kong's locality? Would I side with the environmentalists' opposition to building a manicured landscape for elites in a country park, or would I instead advocate the cause of villagers who asserted indigenous land rights? These positions were starkly opposed, and yet they shared a common apprehension of the Sha Lo Tung incident as an instance of cultural and political conflict. That is, both took Sha Lo Tung as an event that revealed the extent to which environmental politics in Hong Kong were inevitably entangled in questions of cultural difference. The only question worth posing, it seemed, was which side one would choose.

It wasn't difficult to guess the right response in different situations. Janet had invoked the Sha Lo Tung incident as an illustration of how tenaciously cultural barriers impede environmentalism in Hong Kong. The villagers' anger exemplified the resentments that she believed were directed toward environmental work. Her story was offered as a tragedy, and the appropriate response, I surmised, was a sympathetic gasp. Many friends and family who had grown up as Cantonese subjects of colonial rule, meanwhile, told the story of Sha Lo Tung with wry smiles. Even if they themselves might have preferred that the valley remain undeveloped, Sha Lo Tung offered a glimpse of local agency, a momentarily effective resistance to environmental NGOs whose most visible representatives were either expatriates like Janet or elites who had been educated overseas. Viewed in this light, environmentalist values derived from origins and commitments beyond Hong Kong's borders.

Still, it was hard to respond properly. Alongside the question of my cultural and political alignment—or embedded in it—were analytic and descriptive questions: What would I take the Sha Lo Tung event to exemplify, and what details of the situation would I highlight to draw Sha Lo Tung into relation with other cases? Would my response draw the Sha Lo Tung Valley into comparison with other natural landscapes and other precious centers of biodiversity? The ecologists arguing for Sha Lo Tung's status as an ecologically significant site were doing this. Would I draw the coalition of NGOs into comparison with other environmental groups and whistleblowers that had challenged governmental oversight in other places, as NGO spokespeople tended to do; or would I emphasize their similarity to transnational environmental coalitions that had historically undervalued local concerns and needs in their single-minded quest to preserve green space? Would I compare the villagers' efforts to develop their land as they saw fit with struggles for indigenous sovereignty, or with the efforts of developers to evade environmental regulation?

All of these comparisons were possible. They also felt familiar, not unlike the frameworks I had learned during graduate school to gain some analytic distance, or a metaview, of environmental science and environmental politics. And therein lay the problem, I realized, as I sat open-mouthed while

Janet and others waited for my response. The tools that I relied on for taking a step back from controversy in fact located me in it.

Knowledge Practices in Environmental Politics

This book is about knowledge practices in environmental politics and anthropology. It offers an ethnography of the knowledge practices and cultural political negotiations that underlay a surge of environmentalist activity in Hong Kong in the late 1990s, just before and after the region's handover from British to Chinese sovereignty. I describe some of the details, informal occurrences, and biographic and cultural idiosyncrasies of Hong Kong and its environmental arenas, partly because I think they are important as empirical reminders of how environmentalism as a global phenomenon is constituted precisely through such contingencies, and partly to convey how much the terms of debate in Hong Kong's environmental controversies anticipate our terms of analysis—just as they did in Sha Lo Tung. Engaging in this ethnographic work thus prompts me to consider the limits of certain analytic habits leaned upon by anthropologists, social theorists, and others who are thinking through the messy scales of politics in transnational and transcultural situations.

In what follows, you will read about a mobilization to preserve a species of white dolphin (*Sousa chinensis*) that became the mascot for the Hong Kong handover in 1997; a struggle to protect an aging fishing village from redevelopment; a tale of orchid kinship; the negotiation of "locally appropriate" expertise in an NGO-village collaboration; environmentalists' narratives of environmental awakening; and efforts to substantiate air quality as an object of political and medical concern. As the book moves through these moments, I will ask you to attend to their techniques and politics of specification, exemplification, and comparison.

Specificity, example, comparison: these are words I rely heavily upon in this book. I hope you will read them not as self-evident concepts but as trigger words inviting reflection on the practices of specification, exemplification, and comparison through which such concepts are made real in the world. These words are meant to direct our attention to the ways in which Hong Kong environmental practices draw and conceptualize connections between places, between species and other species, between forms of life and their environs, between what is considered big and what is considered small, between particulars and universals, between particular cases of a common rule, between specificities and generalizations, between grounded details and ambitious abstractions.

You might read this book as an ethnography of comparison. It offers a close view into some of the comparisons, differentiations, and articulations that characterized environmental, political, and social scientific knowledge production in Hong Kong in the late 1990s, and it analyzes the forms these acts of relation-drawing took. Its objective is not simply to offer an anthropological diagnosis of Hong Kong or the environmental problems faced there, but to see whether thinking through these politicized comparative acts might help us develop a method for bypassing certain quandaries encountered by scholars trying to theorize politics on a global scale. In particular, I am interested in the roles of universality and specificity in political mobilization and the possibilities they hold for collaboration through and across difference.

Thus, while not a comparative study in the usual sense—I will not draw comparisons, for instance, between environmental arenas in Hong Kong and those in other parts of the world—in its way, this book is decidedly comparative. It juxtaposes and draws relations between the kinds of comparative acts I witnessed among activists, scientists, laypeople, and others in Hong Kong and the kinds of comparative acts used by anthropologists, social theorists, and political philosophers to make sense of the world. In other words, it compares comparisons.

Writing this book from within the discipline of anthropology, I cannot help thinking of Franz Boas, the nineteenth-century anthropologist who famously repudiated the discipline's "comparative method" of distinguishing and locating cultures in a racially segregating evolutionary ladder.[2] Boas, we should remember, never ceased to think comparatively; he simply demanded that we jettison the assumption that comparative differences denote differences in degree of human evolution. In other words, he cautioned us about the frame in which anthropologists emplotted the various cultural specificities they encountered in their work. A century later, I wonder how Boas would respond when American colleagues express surprise that I would study environmental politics in Hong Kong, ask whether environmental politics have caught on "yet," or seek insight into why certain places are more "environmentally backward" than others. Comparison, my time-traveling companion might remind me, is not in itself the problem. But we need to make explicit the stakes and politics that attend particular lines of comparative thinking, bound up in the very concepts and scales through which, seeing an example, we think to draw a comparison.[3]

Ecologies of Comparison

From the get-go, environmental politics in Hong Kong have been thick with comparisons, explicit and implicit. The first of Hong Kong's environmental advocacy organizations, or "green groups" as they are called after the British fashion, was the Conservancy Association (CA), which opened its office in the colony in the late 1960s with a mission to act as a government watchdog. Little change occurred in the environmental NGO landscape in the following years until the 1980s, when there was a proliferation of new environmental groups. In 1981 the World Wildlife Fund opened a Hong Kong office; a few years later, Friends of the Earth followed suit.[4] Soon after, several smaller groups without international affiliations also emerged, such as Green Power, established in 1988 with a charter to promote green lifestyles, and Green Lantau and Green Peng Chau, conservation groups on two of Hong Kong's outlying islands.

HONG KONG'S HISTORY OF RELATION

Until 1841 the Cantonese phrase *hong kong* referred only to a tiny harbor on what Lord Palmerston, Queen Victoria's foreign secretary, famously called a "barren rock" of an island off the southern coast of imperial China. The literal meaning of the phrase, fragrant harbor, made it an apt signifier for that small cove at the center of the incense trade. How the harbor's name came to represent the entire island is unclear. Was it a translation error? Did British traders and officials misunderstand their Chinese trading partners and think the Cantonese words referred to the island as a whole? Or was it perhaps that merchants and traders on both sides of the linguistic divide made the synecdochic slip knowingly, for the sake of convenience, as the island became more important to their mutually profitable exchanges? In any event, "Hong Kong" became an internationally recognized piece of discrete geography through a history of interactions and violences between British and Chinese traders, officials, and settlers.

And this geo-body grew. Hong Kong, the island, was ceded as a colony to the

British in 1842 after the first Opium War. Then, in little more than half a century, the appellation "Hong Kong" came to encompass not only Hong Kong Island but also the Kowloon Peninsula, which lies just to the north, as well as lands that would come to be known as Hong Kong's "New Territories." Some scholars have called this a semicolonial period for China, with intense European jockeying for commercial footholds in the Middle Kingdom, which until the late nineteenth century had been notoriously indifferent to European goods. China was never occupied outright, but it did yield the islands of Hong Kong and Macau to British and Portuguese colonization, and the British seized the Kowloon Peninsula in 1861 after the second Opium War.

Thirty years later Hong Kong grew again. Germany and Russia had taken the ports of Jiaozhouwan and Lüshun-Dalian by force in 1897, and were granted authority there by the Qing government in 1898. France demanded the cession of a bay in Guangzhou in 1898. British officials, citing the need for greater commercial and military security, subsequently demanded that a larger mass of Chinese land be granted to Britain on long-term lease, if not outright cession. A ninety-nine-year "lease" of the New Territories was subsequently secured in 1898.

Today the island that took the name of its harbor makes up only 7 percent of the region known as Hong Kong, and Kowloon another 2 percent. The New Territories form the bulk of the territory—some 91 percent—comprising all the land between Kowloon and the Chinese border, as well as all the islands neighboring Hong Kong Island.

There are so many dates already. Two more—1984 and 1997—are necessary for understanding the politics at play in this book. In 1984 Margaret Thatcher, the British prime minister, and Deng Xiao Ping, the Chinese premier, signed the Sino-British Joint Declaration. In the declaration the British government agreed to return to Chinese rule, upon expiration of the ninety-nine-year lease of the New Territories, not only the New Territories but the rest of Hong Kong as well. Subsequently, on July 1, 1997, with much fanfare, Hong Kong was handed over from British to Chinese sovereignty. According to the Basic Law drafted in 1986, even if Hong Kong would ultimately be absorbed into the People's Republic, it could be assured of its political and economic autonomy for fifty years. The phrase coined to capture this unique period of political transition was "one country, two systems." Hong Kong would be both common and distinct: a part of the motherland, yet one with its own political and economic systems.

There is a striking uniformity to the names these groups adopted, but against the repeated green backdrop we also see marks of distinction, particularly in the scales of authority and claim. Groups charted to work for causes as broad as "nature," "conservancy," and "earth" vie and coexist with groups mobilized for localities like Peng Chau and Lantau. Already one can see the environmental political landscape marked by crosscutting and competing claims for different scopes—planetary universals and place-based specificities—of conceiving environmental problems and environmental activism.[5]

Accompanying this period of environmental NGO growth was the burgeoning question of environmentalism's place. Were these imported politics? Was environmentalism "Western"? The issue received its most explicit address in the public lifestyle politics of Dr. Simon Chau, a cofounder of the group Green Power, who also garnered headlines and feature articles in Hong Kong's English- and Chinese-language newspapers for exemplifying an environmentally conscious lifestyle. He rode his bicycle to work even on hot, humid summer days. He founded the Vegetarian Eating Society of Hong Kong. And, significantly, he gave frequent speeches in which he traced environmental concerns to the tenets of Buddhism. Chau claimed environmentalism performatively as something fundamentally tied to a civilizationally and religiously figured "Asianness," pulling it away from its perceived belonging to a sphere of "Western" values. In light of this effort to make environmentalism span "Western" and "Asian" cultures, and thereby appropriate to and for Hong Kong, Chau's translingual vocation—he was a professor of translation studies at Hong Kong Baptist University—seems almost too fitting.

Perhaps the most prominent event in the history of grassroots environmental protest in Hong Kong was a wave of antinuclear demonstrations that occurred in the mid-1980s when the Chinese government and Hong Kong's state-run China Light and Power Company jointly proposed to build a nuclear power plant in China's Guangdong province. The plant was to be built in Daya Bay, a cove located in Chinese territory but only fifty kilometers northeast of Hong Kong. Mass mobilizations ensued, in which Hong Kong's relatively young environmental NGOs joined with veteran unions, lobbying organizations, and residents to stand against both the Chinese government and the British-run colonial government of Hong Kong. The protests reached their apogee when organizers submitted a petition signed by one million Hong Kong residents. Both colonial and Chinese governments maintained, however, that the nuclear plant was safe, and construction proceeded apace. The coalition lost.

What strikes me most is how the Daya Bay controversy not only condensed fears about nuclear energy but crystallized skepticism of both current British colonial governance and impending Chinese rule. On Kwok Lai, a professor of policy studies at Kwansei Gakuin University suggests that it was particularly potent because of its resonance with the Chernobyl accident of 1986, and also because "it developed under the shadow of the uncertainty of Hong Kong's political future."[6] The Daya Bay plan emerged only two years after the Sino-British Joint Declaration of 1984, which promised Hong Kong's return to Chinese rule. In addition, the "one country, two systems" Basic Law had just been drafted in what many critics in Hong Kong considered a secret meeting between British and Chinese officials. This was a time when many people in Hong Kong wondered openly whether either power—Britain or China—had Hong Kong's interests at heart, and when the notion of an autonomous political and cultural identity was just beginning to be raised and espoused in Hong Kong.

As a colonial port city and global financial hub, Hong Kong traffics in clichés as much as it does goods and labor. The place often appears over-determined as an icon of "East-West" connection, whether in nationalist Chinese histories, in business magazines, or in its own tourist literature. There is no way to study Hong Kong on its own terms without recognizing that an essential part of Hong Kong's specific character has to do with the ways in which its residents negotiate various discourses of cultural and material exchange, articulated in idioms of mixture, imperialism, colonialism, anticolonial Chinese nationalism, capitalist advantage, and so forth. Hong Kong cannot avoid these historical legacies. Any discussion of Hong Kong inevitably slides into some kind of discussion of East-West relationships, in which "China" stands for the "East" and Hong Kong becomes the hyphen whose function is to differentiate, even as it gathers and conjoins, East and West. Yet how these two are defined, and how their legacies will settle in the middle term, is not completely determined. It is in the work of defining that meaning that an autonomy for Hong Kong becomes thinkable.

The Daya Bay mobilization was shaped by these emergent political questions. The environmental controversy, more crucially, was central to the formation of a conception of Hong Kong as agonistically pitted against both Britain and China. Neither sovereign power, the one million signatories effectively stated, had the interests of Hong Kong's residents in mind.

Congenitally commingled in these politics of environment are not simply "the environment," "nature," or even "cultural conflict" in the abstract but reckonings with the origins of environmental ethics, the appropriateness and scope of environmental consciousness, and Hong Kongers' questioning of both colonial and postcolonial rule.[7] These are messy political matters, shot through with questions about Hong Kong's place in the world. There is no way to imagine a pure nature unmediated by politics. There is also no avoiding questions about other places when asking what should happen in Hong Kong. There is no avoiding the question of specificity, the task of asking what relation Hong Kong's specific problems have to other natures, economies, and politics. There is no avoiding these things because, in fact, ecological politics work precisely through such questions and comparisons, acts that recast the relations—of nature, culture, politics, and more—through which a given animal, plant, health problem, landscape, or question comes to matter epistemically and politically. These gatherings of material relations and their conceptual drawings and redrawings are Hong Kong's *ecologies of comparison*—forms of political thinking and action that are both enabling of and enabled by the problems of environmentalism in the postcolony.

"Ecologies of comparison" is a heuristic I use to keep several things in view at once when thinking about these political renderings of relations; it is less a theoretical framework than a mode of attention I wish to sustain through the course of this book. Its usefulness for me hinges in part on three different senses gathered by the term "ecology." First, there is the sense in which "ecology" works roughly as a synonym for "environmentalist," as in the "ecology movement." A number of ecologies, different moments and forms of environmental politics, interest me here. Second, "ecology" names a relatively young branch of the life sciences concerned with organisms and their interactions with their environs, a broadly construed discipline of study including such subdisciplines as evolutionary ecology, conservation and population biology, and ecosystem ecology. I hold these first two senses of the term together to flag the tight link between scientific knowledge production and politics in environmental arenas. Environmentalist politics are inseparable from the practice and development of environmental science, whether arguing for the importance of protecting particular plants, animals, or landscapes, or proving the reality and implications of global climate change. Finally, "ecology" is used to denote an emergent web of relationships among constitutive and constituting parts, such as when one shifts attention from a particular organism to the entire ecology of which it is a part. This third sense, where ecology glosses roughly as "ecosystem," is, again, inseparable from the other two.[8] The relations among and between different forms of life are not simply "out there" to be discovered, nor are their spatial and temporal scales self-evident. They all must be posited and established through scientific (ecological) research.

I qualify the ecologies discussed in the book as "ecologies of comparison" because, in addition to highlighting the inseparability of political and epistemic practices in environmental arenas, I want to think about how the political knowledge practices emergent in these arenas rely on methods of comparison to call relations of interdependence, connection, and disjunction into being. Furthermore, I wish to address how the knowledge practices constituting environmental politics rely upon and generate scales of comparative analysis—local, global, specific, general, particular, universal, species, ecosystem—in the course of drawing ecological comparisons and relations. These ecologies of comparison in turn attribute significance to specific objects of ecological concern. Ecologies work through comparisons, and comparisons work through ecologies.[9]

These ideas should become clearer as I give the examples behind them; for now let me say that I am interested in modes of thinking and acting in environmental politics that attribute specificity and value to a form of life, assert its difference from other forms, and locate it within broader fields. Such conceptualizing practices are important to think about, for they shaped not only environmental knowledge production but nearly all cultural and political knowledge production in Hong Kong in the years surrounding the region's handover. It was precisely through such conceptual practices that people came to know and to politicize as "endangered" not only certain species of flora and fauna but also other Hong Kong "specificities," such as state autonomy, cultural identity, and global economic niche.

The politics of endangerment and specificity are addressed most explicitly in chapters 2 and 3 of this book. In chapter 2 I ask how we come to know and to care about certain forms of life as endangered, doing so through an ethnographic look at the knowledge practices that accompanied two prominent environmental mobilizations of the late-1990s. The first of these mobilizations was an effort to protect a population of pinkish-white dolphins; the second was a struggle against government plans to redevelop a fishing village that was in economic decline. Population biologists and amateur folklorists worked as salvage scientists in these mobilizations, rigorously documenting forms of life on the brink of extinction, establishing dolphin and village as endemic and endangered species and culture. I suggest that an anticipatory nostalgia inheres in these political knowledge practices and explore the political dangers and possibilities this nostalgia offers. Endangerment imbues lives at once with specificity and fragility, casting them as quickly becoming out of joint with their milieus.

Chapter 3 pays closer attention to the practice of specification inherent in the politics of endangerment and reflects on how we might think about the context of specification in Hong Kong. In particular, I'm interested in how we connect scientists' penchant for discovering unique Hong Kong species and Hong Kong's transition from British to Chinese governance. Our ethnographic starting points here are plant biologists studying *Spiranthes hongkongensis*, a novel orchid species, and anthropologists studying Hong Kong culture, politics, and economics. These scientists and social scientists were collaborators, I suggest, in realizing a concept of life as specific, ecological life. The Hong Kong orchid was like, but different from, a Chinese orchid, just as Hong Kong culture was like, but distinct from, Chinese culture. These specifications, and the political claims they enabled, all hinged on systematic techniques of comparison—visual, genetic, demographic, ethnographic—and they resonated with concerns voiced about Hong Kong's political autonomy from China after the handover. How do we interpret this resonance, or its political import? To throw these questions into relief, I consider them against the foil of a moment in eighteenth-century Germany when scientific and political writings similarly came to share a conception of life—not as specific, however, but vital.

Specificity at times manifests in environmental controversies through calls for "local appropriateness." These are calls to guard against the unreflective "universalization" of environmentalism, against the making of environmentalist demands across different contexts in cookie-cutter fashion. They also fit remarkably well with the convention in American cultural anthropology of thematizing the importance of contextual specificities. If parties in environmental controversy themselves voice critiques of universality, what does this suggest about the usefulness of academic arguments attempting to do the same? This is the question guiding my discussion in chapter 4, of an environmentalist coalition of indigenous villagers and Greenpeace activists brought together to fight against the construction of a municipal waste incinerator in Hong Kong's New Territories. In a situation where neither scientific expertise nor local knowledge had political efficacy on its own, villagers and activists generated a form of political counterknowledge that hedged against the declining values of universality and particularity by staging and articulating both. Looking closely at a moment of translation during a town meeting, I suggest that rather than interpreting some parties in environmental controversy as universalizing and others as grounded in a particular context, we must reckon with how the terms we use to think about scales and locations are themselves produced self-consciously in environmental action.

If chapter 4 thematizes translation, chapter 5 focuses on some of the translators, who themselves grapple with questions of location on a more subjective level. Grounded in stories of environmental awakening, travel and cosmopolitan belonging, and principled action, this chapter illustrates what it meant for some subjects in Hong Kong to be faithfully environmentalist. At the heart of environmental vocations in Hong Kong, these stories suggest, were ongoing practices of self-care and self-comparison within local and translocal ecologies of gender and expertise. Through these practices, environmentalists continually recast the terms of relation between themselves and other Hong Kongers, as well as between Hong Kong and other places. The environmental cosmopolitanisms that appear here, and the environmental ethics they foster, emerge through well-traveled routes and commitments to place, rather than commitments to a world or planet in general.

The final chapter, "Air's Substantiations," turns to Hong Kong's air and some of the acts and practices through which air has become a meaningful, knowable, and eventful substance in Hong Kong—as a medical fact, as a bodily engagement, as an international index, and as a medium of social difference. I ask how the varied and shifting materialities of air—as it moves across the boundaries between regions, breathing bodies and their environs, and different domains of experience—might inform our thinking about what counts as solid argument, in environmental as well as other arenas. The chapter works mimetically, gathering disparate accounts of the particulate-measuring machines, doctors, and other actors that condense aspects and experiences of Hong Kong's air to make concerns about air quality more epistemically weighty. In so doing, it suggests rethinking particularity and universality through a language more attuned to material movements between diffuseness and solidity, abstraction and concreteness.

Each chapter examines exemplarity and comparison, and each itself works through exemplification and comparison. A key theme in this book is that environmental politics and knowledge production in Hong Kong do not work through the application of universal principles, or through a universalization of particulars, but rather through ongoing oscillations between detail and broader claim that ultimately blur the distinction between example and abstraction. To track and illustrate this phenomenon, I pursue the same oscillation as ethnographic method. Each of the chapters thus oscillates between the details of a particular problem of Hong Kong environmentalism and a problem of theory until the example's details blend into the language and practice of concept building.

As you may have noticed, the relation between "environmentalism" and "Hong Kong politics" assumes a variety of forms in this book. At one moment one appears to mediate the other—as if, for instance, the ways people in Hong Kong came to grapple with particular problems of the environment refracted other social, political, and economic questions. At another moment their coincidence looks more like a conjunctural articulation—a contingent yet compelling coming-together of specific political questions and techniques. Noting homologies between the environmental as such and the political as such seems to demonstrate a discursive resonance, conveying an epistemic tendency; or it comes across as provocative analogy. Complicating things further, at some points in this book I create or draw the relation; at other times, other people do the relating, such as when environmental activists smirk at the Hong Kong government's choice of a threatened dolphin for a political mascot. I illustrate and track different figurations of this relation in action, rather than arguing that it is best characterized in one particular way. I do this for two reasons: first, to maintain that multiple relations hold between "environmentalism" and "Hong Kong politics" (to the extent that we can consider them separate in the first place); and second, to emphasize that the way people drew both the two and their relations was itself a self-conscious form of political critique. These various renderings—that is, the ways in which one drew "environmentalism" and "Hong Kong politics" together—were the possibilities of Hong Kong's "ecologies of comparison."

These political and conceptual ecologies lived in a stacked ethnographic present. I conducted most of the field research for this book in the late 1990s: a fifteen-month stay beginning in September 1999, two shorter stays in the summers of 1997 and 2004, as well as family visits throughout my life. The ethnographic present of this book could thus be interpreted as meaning the late 1990s. It will become clear, however, that this present, like most presents, folded into itself other dates and other times, not to mention other places: key decisions in the 1980s, the handover of 1997, the origins of

Hong Kong culture, as well as unknown ecological, political, and economic futures. (See "Hong Kong's History of Relation" sidebar.) Hong Kong, already overdetermined as being in transition from British colony to specially administered Chinese city,[10] was also gripped by a regional financial downturn that followed the fall of the Thai bhat, a situation that offered its own idioms of precarity and forecast. Thus histories and counterhistories, as well as futures — both ones hoped-for and others worried-over — tucked unavoidably into people's sense of what was here and now. The ecologies of comparison tracked in this book took part in this disjunctive present-making, activating both spatial and temporal relocation, and reflecting on what might transpire in other places and in future times. It is the layered nature of what might constitute and count in and as "the present" that enables some of the affective politics discussed here.[11]

Specificity and Method

Specificity was not always a topic for this book. Initially, it was simply its method. I envisioned writing a book that would convey the importance of reckoning with regional, cultural, and political specificities even when — or precisely when — working in environmentalist arenas. I wanted an account of environmental politics that would convey the singular history of Hong Kong, and show the ways in which different threads of that history coalesced into particular junctures of environmental controversy. To explain, let me return one last time to the example of Sha Lo Tung. I found the story compelling, for it tangibly illustrated the specific kinds of messiness one grapples with when doing environmental politics in Hong Kong. Its details pointed me to global histories of environmentalism as well as to singularly regional histories: classifications of indigeneity and land laws dating to the beginning of British colonization; a country parks ordinance sponsored in 1976 by a British governor who fancied hiking; the significance of speculative real estate ventures in the Hong Kong of the 1990s; the ease with which British expatriates lived and worked in the colony; the crucial development of ecological sciences and conservation biology alongside political environmentalism; and a particular history of environmental activism in Hong Kong that rendered it impossible to conceive of doing environmental politics without raising questions of their cultural provenance and specificity. Sha Lo Tung was a perfect example of the specificity I wanted to convey.

When I thought more about the efforts by ecologists to argue that the Sha

Lo Tung Valley was a Site of Special Scientific Interest, however, I realized something. We were doing versions of the same thing: generating value—for Sha Lo Tung and for Hong Kong—by producing knowledge about and arguing for their place-based uniqueness. While I had come with the intention of representing local specificity or historically specific conditions of possibility and contextually specific assemblages of environmental expertise, others were already embroiled in precisely this task. People in a variety of contexts in Hong Kong readily and constantly explained to me the specificity and importance of different places and forms of life, and it was impossible to ignore the fact that these explanations were interventions shaping what comparisons one could make analytically and the stances one could adopt politically. Residents of a fishing village spoke easily with me about the historical and cultural uniqueness of their village and customs, as did inhabitants of a relatively well-off indigenous village in Hong Kong's Northwest New Territories who hoped to halt construction of an incinerator nearby. Biologists argued that species of dolphins and frogs were endemic to the region. Environmental consultants emphasized the local expertise they gained from regional work.

People also spoke readily of distinguishing characteristics of Hong Kong as a whole: businesspeople celebrated Hong Kong's global niche as a "gateway to China"; academics wrote of uniquely hybrid cultural production; and government officials declared the importance of a "locally appropriate" environmentalism. When colleagues at the Centre of Asian Studies at the University of Hong Kong—my institutional base during my field research—inquired collegially about my progress, my reports of efforts by different people to negotiate an environmentalism grounded in Hong Kong's particular circumstances surprised no one.

I began to realize that any analysis of the specificity of environmental activity in Hong Kong must take into account the travels, forms, and transformations of the very notion of specificity as it manifested in such activities. In attending to the overlaps between discourses of endangerment around natural species and those around cultural specificity, I began to wonder: How had the notion of specificity—and the knowledge-making projects of substantiating specificity—come to be so important for Hong Kongers and for cultural analysts such as myself, and what political work did it do? How had these ethnographic-analytic and political desires for specificity come to overlap in Hong Kong? What kinds of specificity were at stake? If nothing else, what became clear in my conversations with government officials, village residents, activists, and consultants, all of whom had emphasized the importance of local specificity even while advocating divergent political projects, was that one could not speak of specificity in a general way.

As I traveled among different sites I began not only to document the expert and lay production of Hong Kong's biological, cultural, and political specificities, but also to compare the emergent logics and narrative forms that made different types of specificity in Hong Kong (such as species, culture, locality, and state autonomy) possible and meaningful in the first place. I gradually came to understand these logics and forms as *ecologies of comparison*, conceptual practices through which a given event or form of life came to matter—in environmentalist and other political terms.

This book is about environmentalism, and it is about how we think about environmentalism. It is about Hong Kong and how we think about Hong Kong. It is also about how we think about environmentalism in other places, and how politics, freedom, culture, and expertise are thought and practiced in worlds of comparison and relation making.

This book is also about (and for) some wonderful, earnest, smart, critical, hopeful, committed people who try to change the world: to save nature, preserve culture, defeat capitalism, safeguard health. These people are dear to me; they and their brave acts inspire hope. I often desired to be useful to them while conducting my research—to make relevant, actionable knowledge for friends who were activists in Hong Kong as well as for local scholars.[12] For better or worse, however, this was not the path I took in researching and writing this book. Its shape derives not from urgency, but from mulling over the urgency and shape of some political acts in Hong Kong, and the urgency and shape of certain ways of theorizing culture and politics. Its method has been to take note of what gets left out, an ongoing practice of reframing and revising things I have done, said, thought, or written. This method has made me slow to speak at times, but it guards a hope that building new knowledge practices rather than relying on old ones can open different political horizons, for friends in Hong Kong and for others. Too much mulling for an activist handbook, perhaps—but a good amount, I hope, for the practice of conceiving other possibilities.

Wing Hung had already left for the day by the time I arrived at Greenpeace's small office in Hong Kong's Sheung Wan district, but I was happy to run into Maria, the organization's press officer. We were making some of our typical small talk and arranging to meet for dinner when I noticed a familiar twinkle in Maria's eye—the kind that said a good story was on the way. She tilted her head with a smile toward Wing Hung's desk, which I now saw was strewn with junk food wrappers, and with obvious relish began to tell me how the wrappers had gotten there.

Wing Hung had devised a plan to advance a campaign he was organizing against genetically engineered (GE) foods. Figuring that people would be especially sensitive to learning about strange ingredients in popular snack foods, Wing Hung walked to the Wellcome grocery market nearby to procure samples that could be tested for GE ingredients. Some of the items he picked randomly; but he targeted others, like candy bars made by Nestle, because their manufacturers had a history of using GE ingredients. If these popular foods contained genetically modified components, Wing Hung knew, the Hong Kong public would be incensed.

All in all, a brilliant plan. But Wing Hung encountered

an unforeseen challenge at Wellcome. It turned out that the items he wanted were on promotion—two for the price of one! It would have been mad, he thought, to pass up a deal, so he returned to the office with the freebies. After mailing his samples to the Austrian laboratory that had agreed to test the foods, however, Wing Hung found himself with a tableful of duplicate snacks. Not one to waste food, even of dubious genetic origin, he ate his way through the mountain of chips and candy bars while working at his desk for the rest of the day. Maria giggled, "He said he really loves his work." Indeed, every time the incident came up afterward, Wing Hung would beam, "I'm passionate about my work! I'm so passionate about it that I ate it all!"

Wing Hung had three main passions. The first was for activism. As a student at the Chinese University of Hong Kong, he engaged tirelessly in student politics, organizing and participating in campaigns aimed to improve the living conditions of Hong Kong's urban and rural poor. His passion was contagious: he brought me to my first political action in Hong Kong in the summer of 1997, a protest and march organized by the Hong Kong Student Union about rising rates of unemployment in the city. He also invited me to a workshop about the political and economic implications for Hong Kong workers of an Asian financial crisis.

Wing Hung's second passion, related to the first, was for critiques of globalization and development. He was well-read in most types of social theory, but his loyalties clearly leaned toward the explicitly Marxist—toward the works of world systems and globalization theorists. These loyalties were formed during his days as a student activist at Chinese University, and they were solidified by his studies with the Marxist historian Arif Dirlik, who taught for a year at the University of Science and Technology in Hong Kong, where Wing Hung had begun work on a master's degree. Wing Hung's passion for globalization theory was so intense, in fact, that his friends gave him a nickname: Wing-Kauh. The Cantonese word *kauh* means "sphere" and is a component character in the Cantonese term *chyuhnkauhjyuyih* (globalism; literally "whole sphere-ism"). Wing Hung's alias, then, implied Wing-Global, or Wing-Ball, a name made funnier by the reference it made to Wing Hung's roundness. "This is the globe, I am the globe! I am the true face of globalism!" he roared, clutching his belly.

The belly Wing Hung attributed to his third passion: good food. He made sure that life was not all business. When we first met in the summer of 1998, Wing Hung was a full-time graduate student with summer vacations off, and his girlfriend, Siuming, along with two of their close friends, Faanshu and

Ginny, had just completed their undergraduate degrees at Chinese University. The five of us had plenty of time for fun, and the group made sure my Hong Kong education was both well-fortified and well-rounded. One evening after a typically gut-busting dinner, Wing Hung sat me down to map out the four main curse words of Cantonese on a paper napkin, while Faanshu, Ginny, and Siuming carefully explained the grammatical rules for inserting vulgarities into words, all of them applauding when I generated novel argot of my own.

As boisterous as he could be, however, Wing Hung seemed quiet, almost bashful, when he told me about his new job in 1999. In many ways, Greenpeace epitomized for him both environmentalism and globalism, universalisms whose blind spots Wing Hung had always been quick to critique. The early days of our friendship were marked by passionate discussions of various critiques of development, and we had agreed that a danger of cultural imperialism and developmentalism lurked in "global environmentalism." "Now you can do your research about Greenpeace," Wing Hung said as he welcomed me back, "and I'll tell you about all the critiques."

Greenpeace's executive director, Chan Yiu Kwong, once told me that it had been difficult to recruit Wing Hung to the organization. Yiu Kwong was promoted to executive director from his position as a Greenpeace campaigner in 1999, and his first task was to complete the rest of Greenpeace International's (GPI) "localization plan," which mandates that all new Greenpeace offices hire a "local" executive director within three years, and a local staff within five. Yiu Kwong was perhaps ideally suited to this task of "localization"; he knew Hong Kong's particular social and political problems well from his background as a social worker. The minimum requirement for staff, according to GPI's plan, was an ability to speak Cantonese or Mandarin, but Yiu Kwong wanted more: he wanted to staff Greenpeace with campaigners who cared about and were versed in Hong Kong politics.

Wing Hung seemed to Yiu Kwong a perfect fit for Greenpeace. He showed obvious capacity for quick thinking and research, had experience in activism in Hong Kong, and had just begun a leave from his graduate studies. But Wing Hung had been reluctant, wary of implementing an NGO's global vision, of brokering its worldwide expansion. Yiu Kwong finally managed to persuade Wing Hung by arguing that Greenpeace provided the means to accomplish Wing Hung's goals. Wing Hung could use Greenpeace's international clout and its far-reaching networks to address the anticapitalist concerns he held most dear. And the organization, with Wing Hung's help, could be a different kind of Greenpeace. Those aspects of Greenpeace's environmental politics concerned strictly with nature, ecology, or building a global environmental movement may not have compelled Wing Hung, but the possibilities Greenpeace offered to take on large corporations did.

Wing Hung's junk food binge seems contradictory at first; he ate what he argued against. For Wing Hung, though, the problem with genetically engineered food lay not in the food itself, but in the conditions of its production. His motivations had little to do with concerns about taste or safety. While uncertainties about environmental harms and health rationalized the campaign on environmental grounds, more important for Wing Hung was what was certain. It was certain that those most ready to profit from the ascendance of GE food were multinational corporations in the food production industry, such as Monsanto, which already monopolized much of the seed industry through its genetically modified seeds. Wing Hung's campaign against the wholesale introduction of genetically modified organisms into Hong Kong's foodways, then, was a battle against corporations—not necessarily a fight for the environment, but a fight against what he saw as the most insidious side of capitalism; not against an unknown unnatural object but a known capitalist one.

Wing Hung's binge thus illuminated the particular space he carved for himself in the arena of environmental activism—one marked not by contradiction but by ambivalence. He cared enough about his campaign to work twelve-hour days with alarming regularity, and at the same time he was not above snacking on the very foods his campaign aimed to vilify. He was planted firmly in an environmental activist project, producing environmentalist discourse daily in the public realm; and yet he allowed himself analytic distance between himself and environmentalism, just as he had as a student. He materialized this distance with a politically perverse intimacy, by incorporating his environmental nemesis.

ENDANGERMENT

First Tale: The Dolphin

In the early 1990s a pink, nearly translucent dolphin began dominating the headlines of Hong Kong's newspapers. Dolphin corpses had started washing up on Hong Kong's beaches with deep gouges in their flesh, and though no one had paid much attention to the pale cetaceans before, the shock of these bloodied bodies—an arresting power bound up in histories of other environmentalist mobilizations of sentiment around marine mammals like whales, otters, and dolphins—vaulted the pink dolphin into novel prominence in the public eye as an urgent environmental issue.[1]

Fault appeared to lie in the pending, controversial construction of a new mega-airport. Approved by the Hong Kong government in 1989, the Airport Core Program was a project of stunning proportions, one best described through superlatives. The airport itself—to be built on Chek Lap Kok, an islet just off of Lantau, Hong Kong's major outlying island—was planned to be the world's largest. It would also become the largest public-works upgrade on earth, leveraging just over HK$21 billion toward building not only the airport but also two suspension bridges, two submerged tunnels crossing Victoria Harbour, and an entire town to hold two hundred thousand people including airport workers. In addition to this, construction would require thirty-four kilometers of new highways and railways. By having the biggest airport in the world, city planners promised and hoped, Hong Kong could be assured of its continued position as Asia's most important hub for business and freight.

But planners faced a number of dilemmas. There was, of course, the extensive infrastructure required to connect the new airport to the rest of the territory, particularly to Hong Kong's downtown and commercial areas. Still larger, though, was a problem of a more practical nature. Over twelve hundred hectares of free land were needed to support the airport complex, but Chek Lap Kok itself measured only a little over three hundred hectares, an area less than 25 percent of what planners required. Furthermore, the island was not only small, it was hilly—not an ideal geography for runways. So how did this 302-hectare island grow to 1,248 hectares? How did Chek Lap Kok quadruple in size?

First, in 1991, the entire island, hills and all, was flattened to six meters above sea level by a major Hong Kong and Japanese construction firm that had won the HK$1.2 billion job from the New Airport Projects Coordination Office (NAPCO). Also leveled was a much smaller island unfortunate enough to be nearby. That done, roughly nine hundred hectares—about nine square kilometers—of flat land were still lacking, so they were filled in with 347 million cubic meters of earth dredged from the bottom of the harbor.

It was this work of growing Chek Lap Kok, of creating land out of sea—what is termed "reclamation" in planning circles—and other aspects of the construction of the airport that environmentalists hypothesized were responsible for the sudden appearance of dolphin corpses on Hong Kong's shores. The dirt and rock that were "reclaimed" had come, after all, from the harbor floor. What resulted, scientists argued, was both a loss of shallow-water bay habitats and a dramatic increase in water murkiness from the siltation, which forced the pink dolphins to relocate. While searching for new places to live, they ran afoul of construction and trade barges.

The pink dolphin maintained its presence as one of Hong Kong's central environmental issues for the next several years. Green groups, environmental scientists, and government officials vigorously debated the importance and uniqueness of the species, and *Sousa chinensis*, or the Chinese white dolphin, eventually earned protections under Hong Kong's Wild Animals Protection Ordinance and Protection of Endangered Species Ordinance. The dolphin's status as the paradigmatic Hong Kong environmental problem was indisputable.

Then on July 1, 1997, something remarkable happened. On that day, the pink dolphin swam into political prominence beyond the environmental arena, gracing stickers and other paraphernalia as the Handover Committee's handpicked mascot for the gala ceremonies marking Hong Kong's transformation from British colonial territory to Chinese Special Administrative Region. Environmentalists found this ironic—that an endangered species should represent Hong Kong as it came under the power of a country notorious for environmental degradation. It did not exactly bode well for Hong Kong, a staff member at one green group joked to me, if China was going to treat Hong Kong as carefully as it did its endangered species.

Second Tale: Tai O

A few years later, a small fishing village on Lantau Island named Tai O made news of its own. In April 2000 the Hong Kong Planning Department published a strategy for "revitalizing" the village. Tai O had been known, up until the 1950s and 1960s, primarily as a dirty backwater; recent years, however, saw a surge of interest in the village. Now tourists from overseas, as well as from Hong Kong, came to Tai O to see the boats, the old fishermen, and the *paang uk*, or stilt houses, as well as to buy the salt fish and shrimp paste that had historically formed the basis of the local economy. The Planning Department wished to capitalize on this popularity and hoped to bring even more tourists to Tai O through its revitalization plan. Already 300,000 people visited each year. With new facilities, planners projected that the village might see another 150,000.

The prospect of such a drastic increase of traffic caused some to take pause, but most shocking to residents and lovers of Tai O was the planning proposal's suggestion that a strip of stilt homes adjacent to Tai O Bridge be demolished. The authors of the proposal reasoned that a rebuilt version of this culturally significant architecture, perhaps with accommodation facilities, would be more attractive to visitors.[2] Then a government official made an unfortunate reference to the good example offered by stilt houses in Malaysia. His statement triggered widespread speculation among Tai O residents and others that the government hoped to build Malaysian-style stilt houses in the place of the ones it would demolish. "They're crazy!" muttered the usually kind-eyed woman who sold umbrellas, water, and canned goods across from the site of the old Tai O market. "They want to take down our paang uk and build new Malaysian ones in their place? What's the point of having Malaysian houses in Tai O?"

No, no, government officials contorted themselves uncomfortably. They did not mean to build Malaysian-style homes; they simply meant to use Malaysia's successful synthesis of traditional stilt architecture and a modern tourism infrastructure as a model for their own growth. Despite their assurances, however, the rumor persisted.

Resistance was vocal. Newspaper columnists inveighed against the plan, and a number of concerned Tai O and Hong Kong residents formed a group to protest the government plan and to develop a counterproposal for the touristic development of Tai O. "People don't come to Tai O to see new things," argued Wong Wai King, a village resident and activist. "They could see new things anywhere. It's the old stilt houses that give Tai O its uniqueness, lets you appreciate the life here. I don't know why the government always wants to take things down and build them from scratch."

"We're fine with tourism," she told me at another time. "Tai O needs it economically. There's no work here for our young people, so everyone leaves. But it doesn't have to be this tourism that the government wants. It should be an *eco-* and *culture*-tourism." It was this ecotourism that had first drawn me to Tai O. I wanted to see how Wong Wai King and her friends progressed in their plans to develop an alternative ecotourism strategy for Tai O, how they produced Tai O's environment and culture as a good-enough commodity to mitigate the need for the government's proposed overhaul. To my eyes, prospects looked good. The newspapers were on Tai O's side, and government proposals to build modern stilt homes in Tai O seemed ridiculous enough to be defeated. The chairperson of the Rural Committee, a notoriously pro-development government body, even reversed his stance and claimed he had never supported the demolition of Tai O's stilt homes. The homes, it seemed, could be saved.

Then, on July 4, a runaway fire in Tai O burned one hundred families' stilt homes to the ground.

• • •

This chapter considers the sciences and politics of *endangerment*. Endangerment is a key trope in environmental politics. It structures images of simultaneous tenuousness, rarity, and value. To speak of an endangered species is to speak of a form of life that threatens to become extinct in the near future; it is to raise the stakes in a controversy so that certain actions carry the consequences of destroying the possibility of life's continued existence. Species can be endangered, as can ecosystems. And, as environmentalists grapple increasingly with the tight bonds that can be formed between people and places, between situated practices and specific landscapes, and between what are commonly glossed as culture and nature, discourses of endangerment have come to structure not only narrowly construed environmental politics, but also politics of cultural survival.

Endangerment is also good to think with in thinking about Hong Kong, though I should take care to be clear about what I mean by this. In the wake of the Sino-British Joint Declaration in 1984, China's deadly crackdown in 1989 on student demonstrators in Tiananmen Square, and an economic downturn in Hong Kong that immediately followed the political handover of 1997, many in Hong Kong and elsewhere suggested that Hong Kong's future might be in peril. When I say that endangerment is good to think with in thinking about Hong Kong, it is not to reproduce this apocalyptic discourse.

Instead, precisely because it seemed such common sense, I want to subject this sentiment to analysis and comparative study. While in Hong Kong in 1998, 1999, and 2000, I could not help but see similarities between the languages of endangerment circulating in the loosely environmental arenas that were my field sites and the anxieties about Hong Kong's future that pervaded local and international media. Both domains focused on the prospect of a drastically changed milieu, and the prospects of a particular object within that milieu—such as a dolphin, a culture, or a capitalist entrepôt—of weathering such drastic changes. Both, too, were idioms in which *uniqueness* mattered intensely. How did these facts of endangerment come to be? Through what processes and practices had what seemed to be a matter of fact—that there was cause for concern after the handover and the Asian financial crisis—come to count and to matter as so much common sense?

This chapter and the next take the first steps toward understanding such processes by considering endangerment practices on a smaller scale. They ask two basic questions: How does endangerment come about? And what are the implications of conceptualizing politics in terms of endangerment? Focusing particularly on two exemplars of endangerment in Hong Kong, the pink dolphin and the fishing village Tai O, I aim to explicate the technical and affective production of endangerment—that is, to elucidate the tropes and techniques that enable certain forms of life to come to matter as objects of knowledge and love, as endangered things. My objective is not only to illuminate the general techniques through which discourses of endangerment function and are produced, but also to sketch out some of the resonances and significances that such discourses had in Hong Kong at the end of the millennium.

This requires some consideration of the time-space of endangerment, or the figuring of temporality and space that endangerment entails. So I also point to the overlaps between endangerment and nostalgia, and suggest ways in which the politics of endangerment, as an *anticipatory nostalgia*, might activate an expansion of nostalgia's political potential.

Sciences of Endangerment

In the interdisciplinary field of science and technology studies, it has been relatively well established that matters of fact are outcomes of knowledge-making practices rather than things that preexist them. Let us direct these insights about science in general toward a more specific question: How is the particular fact of *endangerment* produced? Science studies has shown that certain procedures, such as inscription or entextualization, the mobilization of allies, and chains of translation, are central features of scientific fact-making. As we consider endangerment more closely, the important questions have to do with distinguishing kinds: What *kinds* of facts? What *kinds* of inscriptions? What *kinds* of alliances? What *kinds* of intermediate facts, mobile inscriptions, and enabling alliances are required to stabilize the "endangerment" of an object, or better, to stabilize the "endangered object" as such?

The procedures of producing scientific endangerment are particularly visible in the pink dolphin's case, because the biologists charged with establishing the dolphin's endangerment started almost from scratch. People in Hong Kong had barely noticed the dolphins before, and biologists had never studied the creatures in depth. As two Hong Kong researchers put it, "Although white dolphins have been known to inhabit the waters around Chek Lap Kok for many years, the species attracted little attention, partly because the relevant scientific research expertise has not existed locally."[3]

The dolphins first washed up on Hong Kong shores in 1989. The World Wide Fund for Nature (WWF), at that time the most established green group in Hong Kong, began to lobby the Hong Kong government and was aided in its efforts by local newspapers, which ran dolphins on the front pages for several weeks. Eventually, in 1993, the government supported a visit by a dolphin expert, who recommended that the problem receive further study. In response, the Provisional Airport Authority agreed to make $HK23 million available for research. The funds were given to the Agriculture and Fisheries Department (AFD), which then channeled them to the Swire Institute of Marine Science (SWIMS) at the University of Hong Kong, to be overseen by

Thomas Jefferson, Chris Parsons, and Brian Morton. With the money, Morton funded two graduate students from England to come to Hong Kong for Ph.D. studies on dolphin ecology. The students researched the ecology, population biology, and behavior of S. *chinensis* and submitted reports of their findings to the AFD. The SWIMS researchers estimated in 1996 that 112 of the translucent marine mammals remained alive, a 40 percent drop from the year before. In 1996 Hong Kong's first marine park was established to provide a managed habitat for the dolphin.[4]

More research followed: a study of a possible bubble wall to protect dolphins from noise, assays of blubber and livers to determine whether they contained organochlorines or heavy metals in elevated amounts. New institutions became involved, and in some cases, like that of Dolphinwatch (a group offering dolphin-sighting boat trips), new institutions came into being. Not only dolphins but individual careers and institutions came to be reliant on the actor-networks forged through dolphin science and dolphin activism.

The dolphin issue sparked a series of public reflections on the general paucity of environmental protocols in Hong Kong. Just as certain species are often used by ecologists as indicator species — species whose health indexes the health of their ecosystem — so did the pink dolphin become an indicator of the health of environmentalism itself. There was no baseline database of flora and fauna, said some; the Environmental Impact Assessment process was a sham, said another; there was no locally available expertise, said still others. Sanctuaries are controversial things in themselves, particularly for an animal that is as mobile as a dolphin and depends on the same thing (estuarine fish) for its livelihood as some people (fishermen). In this way, the dolphin spurred an environmental discourse beyond itself, spurred projects larger than itself. The dolphin became good to think with in thinking about environmental politics.

Endangerment operates through two gestures: threat and specification. The first is somewhat obvious; a threat must be identified, a cause of endangerment. The establishment of the fact of threat, though, however crucial, comes secondarily. It was only after several years, for instance, that the SWIMS group began its tests — on organochlorines, on metals, on noise — to determine the threats to the dolphins. Its first, more important step was to generate its object of concern, the subject (the dolphin) that preceded (in a healthy state) the endangering predication.

By *specification*, I mean to call attention to the demand that the biologists establish that the dolphin was not the same as dolphins in other places; to establish, they hoped, that it was its own *species*. The purpose of conducting extensive population studies—to produce documents about the ecology and behavior of S. *chinensis*—was to make the beached animals count as a form of life with some specificity. The SWIMS researchers had to log a vast array of data in formats established through the disciplines of ecology and population biology. They calculated fecundity, fertility, and mortality figures for the dolphins. They photographed them. They estimated population sizes through line-transect surveys. Only in this way could the corpses washing up on the shores count for anything.

Man Si-wai, a local social critic and professor of environmental education at Chinese University of Hong Kong, commented on the situation: "Ecologists are frustrated to witness the perishing of the Chinese White Dolphin in the Hong Kong waters while there seems little that they can do to revert the tragedy. This is because before they can produce foolproof evidence for the uniqueness of these dolphins and for the absolute impossibility for them to "emigrate" to other waters, there is little chance that serious thought would be given on the part of the Government to sanction any project that is expected to cause further degradation to the waters they now live in."[5] Notice what Man points to as necessary evidence: the uniqueness of dolphins and "the absolute impossibility" of emigration. What is needed is proof that the dolphin is *endemic*, that it is of Hong Kong and nowhere else. The cachet of scientifically verified endangerment has everything to do with place, with residence. Dolphins had to prove the uniqueness of their claim to the Hong Kong landscape by failing to appear anywhere else while continuing to dwindle in numbers in Victoria Harbour. This failure to appear elsewhere is called *endemism* in ecological circles. Everything hinges on endemism in the politics of endangerment. If the dolphins could swim somewhere else and survive, there would be no need to stop the dredging and dumping in Hong Kong.

Man entreated her fellow Hong Kong environmentalists to eschew the race for scientific proof. "The tragedy of pursuing this race," she wrote, "lies in that the proof of the uniqueness of the species would not be 'complete' without adequate data on dolphins found in the whole of East Asia being included. And that is not possible—at least not until long after the local dolphins become extinguished." She suggested that environmentalists argue instead "*for the uniqueness of the Hong Kong dolphins on the basis of their cultural and ethical meaning to the community.* In that way, the 'burden of proof' will naturally be on the developers to prove that no harm is done in threatening the survival of these living natural monuments."[6] The Hong Kong landscape—in Man's formulation an aggregate of earth, water, animal, people, and memories—was marked indelibly with dolphin tracks. The threat to the dolphin's future was a threat to Hong Kong itself. It was a threat to something uniquely Hong Kong.

Uniqueness, Dahksik

The issue of local uniqueness concerned more than environmental scientists in the 1990s. While biologists and activists worked to specify the pink dolphin as unique on biological or cultural grounds, a spate of social scientific research posed similar questions about the uniqueness of Hong Kong and the definition of a particular Hong Kong culture. A contrast between two sociological articles can make this point clear.

In 1984 Greg Guldin published an article delineating a seven-type system whereby Cantonese speakers categorized people in Hong Kong.[7] The list included: (1) *Gwongdung yahn* (Cantonese); (2) *Haakga yahn* (Hakka, seen as rural and unsophisticated); (3) *Chiu Jau lou* (Fujianese, Chaozhou, and Hoklo); (4) Fisher People (including the Tanka and Hoklo); (5) *Seunghoi yahn* (Shanghainese) and otherwise *Bak fong* (Northern) people; (6) *Gwailou* (Westerners)[8] and other foreigners; and (7) overseas Chinese.

In 1997 Gordon Mathews added an eighth term to the mix, *Heunggong-yahn*, or "Hong Kong people." Mathews argued that since the late 1960s and early 1970s, a generation of Hong Kong residents, born and raised in Hong Kong, began to develop a sense of autonomous cultural identity distinct from "Chinese." For Mathews, as well as for nearly everyone concerned with tracking and asserting its emergence, this new Hong Kong identity was inseparable from the fact that in 1981, for the first time in Hong Kong's history, over 50 percent of Hong Kong's population was actually born and raised in Hong Kong.[9]

What interests me here is not the veracity of Mathew's findings but rather the fact that the category of "Hong Kong culture" became an important thing to look for in the 1990s, when it had been acceptable a decade earlier to assume that people's self-identities might have nothing to do with Hong Kong. I will say a bit more about this in the next chapter; for now, let me simply flag that Hong Kong culture itself became an object of knowledge as a specificity only recently. If the occasional articles inquiring into Hong Kong culture were not enough to begin stabilizing "Hong Kong culture" as a matter of fact, the cause was certainly helped by the establishment of the Hong Kong Culture and Society Programme at the University of Hong Kong's Centre of Asian Studies. The HKCSP aimed to boost local and international interest in "Hong Kong Studies," providing funding for doctoral students, including me, who were conducting research on Hong Kong and hosting international conferences on the topic of Hong Kong culture.

A term that I heard used regularly to get at the issue of uniqueness was *dahksik*. *Dahksik*, or in Mandarin *tèsè*, is a compound word—a combination of the character *dahk*, "special," and the character *sik*, "color, quality." The term is perhaps most familiar to China scholars for its function in the phrase coined by the Chinese Communist Party, *Zhōngguó tèsè de shèhuìzhǔyì*, commonly translated in English as "socialism [shèhuìzhǔyì] with Chinese [Zhōngguó] characteristics [tèsè]."

I find it more useful, though, to translate *dahksik* as "uniqueness." You might hear it in phrases like "Tai O has its dahksik, but they'll be lost," or "Every place has its dahksik." The dolphin is one of Hong Kong's dahksik. Dahksik functions on occasion as an adjective—while eating a typical Taiwanese noodle dish, for instance, a person in Hong Kong might say, "This is a very dahksik dish." More than a "characteristic" of something, dahksik gives identifying distinction and implies a context where distinction matters.[10]

Wong Wai King, the Tai O resident and activist, was fond of the idea of dahksik. She never failed to tell me what aspects of Tai O were unique, and she always seemed to be looking for ways to document and collect them.

Recording Culture

I am in the lobby of the YWCA Community Center in Tai O, a small room with newspapers draped neatly over rods for residents to come in and read. The lobby is a kind of multipurpose space. On Wednesdays, Janny teaches an English class. Other days, Rosanna might hold her old persons' committee meeting, where they discuss improvements for Tai O. A few weeks ago, I sat here and watched reporters from the *South China Morning Post* interview Mr. Fook—an elderly fisherman, tall, with a full head of white hair—for a special Sunday insert on Tai O.

Today, a broad-shouldered man stands in the lobby, reading the flyers posted on the bulletin board, holding a large video camera at his side like a briefcase. A slight woman in glasses with her hair pulled back, who looks to be in her mid-twenties, waits patiently. I nod to them politely, then sit down and busy myself with my notebook.

A few minutes later, Wong Wai King stomps up the stairs and enters with a wide smile. "Oh, I'm so sorry, have you been waiting long? Tim, when did you get in? Hi, are you Miss Leung? Hallo! I'm Wong Wai King. Nice to meet you. Have you met Tim? Tim is born-and-raised American Chinese. He's working at Hong Kong University now, gathering data for his thesis. He's helping me record Tai O's *haahmshuigo*." We exchange pleasantries and business cards.

• • •

Haahmshuigo (literally "saltwater songs") is the word used to refer to songs sung by Tanka fishermen in Tai O. They are, Wong Wai King tells me, one of Tai O's dahksik. Though few people sing them today, elderly fisherfolk remember singing them at weddings and funerals, as well as in the course of everyday activities. Wong Wai King, as one of her many projects, wanted to record these culturally significant songs. Some of this work she had already done. She showed me a videotape once, a tape commissioned by the Hong Kong Museum of History with footage of her and an elderly woman from Tai O discussing the songs. Prior to the fire, Wong Wai King and I had agreed that the recording of haahmshuigo would be a good project on which to collaborate. She hoped to involve students in the project, so that the young people of Tai O would participate in the recording of their history.

After the Planning Department published its revitalization proposal, news agencies had flocked to Tai O; and after the fire, their interest only grew. Reporters had come to Tai O to interview displaced residents, to ask for their opinions on whether or not they should be allowed to rebuild what were technically "squatter huts." Today, though, they were after something different, and Wong Wai King wanted me to be a part of it. "Can you come on Thursday?" she had suddenly asked while walking me to the bus terminal one day. "The press, Radio Television Hong Kong, are coming—they want to see us *geiluhk* [literally "remembering-recording"] the haahmshuigo."

• • •

While Wong Wai King takes care of her business in the YWCA office, the reporter and I chat. We hit it off. I'm grateful that she doesn't grill me or intimidate me the way Veronica, another reporter from RTHK, does. She doesn't ask me what my story is, what my angle is, or whether I'm done yet. She asks how long I've been coming to Tai O, how I like it. She asks what I think is most special about the place.

Soon Wong Wai King emerges from the air-conditioned office and we're off. It turns out that the reporter and her cameraman have already scouted the perfect spot—a clearing on the banks of one of Tai O's many canals. The lighting is good—no backlight—and the background scenery is nice. We make our way to the canal, picking up two *popos* (respectful argot for referring to elderly women; literally "grandma") along the way. There is Leng Popo, a heavyset woman with a quick smile, and the woman Wong Wai King calls her *baakneuhng*, a term denoting the wife of one's father's older brother. I fetch them some chairs to sit on, and we start preparing for the shoot.

Wong Wai King has a group of young people—mostly high school age— form a semicircle around the two popos. I take up my position next to Wong Wai King while the cameraman sets up his tripod. After checking sound levels, we begin. Wong Wai King takes her notebook out and starts interviewing the popos. I raise my video camera to record Wong Wai King's interview. And the reporters record us.

"How long have you lived in Tai O?" Wong Wai King asks.

"Generations!"

"Did you often sing haahmshuigo?"

"Yes, yes. Of course we sang haahmshuigo."

"When you got married?"

"Yes."

"Are there some songs that you *deui cheung?*" ("Oppose/face-sing," meaning a kind of call-and-response song structure.)

"No, not when we get married."

"When you fish?"

Wong Wai King directs the last question to Leng Popo. Leng Popo seems to have a case of the giggles. "Yes, we sing when we fish." "Can you sing for us?" "No, it's *laanteng ah, laanteng ah,*" she turns her head away, and wipes perspiration from her face with a handkerchief. It's "hard to listen to"; it doesn't sound good.

By and by, though, Leng Popo overcomes her shyness and begins to sing. At first, partly because she sings so softly and partly because she sings in dialect, I have trouble making out her words. But gradually, I recognize that she is singing one of the songs that Wong Wai King has told me about. It is the same song that she transcribed in her book.[11] It is a song of names:

> Do you know which kind of fish is puffed with white powder?
> I know, the Hairtail is puffed with white powder!
> Which type of fish wears gold and silver on his head?
> Yellow Croaker wears gold and silver on his head!
> Which type of fish has its gallbladder removed?
> The Butterfish has its gallbladder removed!
> Which type of fish swims wayward?
> The Cuttlefish swims wayward!
> Which type of fish has only got one eye?
> The Flatfish only has one eye!

I am excited. Excited to have the chance to record this song that is also a taxonomy. The reporters are happy too, happily recording us. I cannot resist—I turn my camera to record the reporters recording us.

• • •

This is the inscription of the object, the analogue of the SWIMS biologists' studies of and writings about the ecology and behavior of the pink dolphins. Both are practices of accumulation—a systematic accumulation of details that characterize a form of life. Like Hong Kong's biodiversity database, compiled by Hong Kong University biologist David Dudgeon, they are entextualizations, translations of material happenings into symbolic, mobile information; they seize culture and nature from the jaws of disappearance. These are the salvage sciences of endangerment.

A few things seem important to note about this moment of recording. Obviously it was a performance—one carefully orchestrated, located, lit, and cast. It was even rehearsed after a fashion; the interviewees were people that Wong Wai King had interviewed before. Though Leng Popo was shy, she knew what Wong Wai King expected of her and had in fact sung that very song for her many times before. This was, in many ways, a replay of the Hong Kong Museum of History video that Wong Wai King had already helped to make.

But the recording at the canal was also something more. For what we performed was not only culture, but the *remembering-recording*, or geiluhk, of culture. (The act of recording is not absent in the Museum of History video, but it is less in the foreground. The cameraman is invisible in that footage. In our newscast, by contrast, both Wong Wai King with her notebook and I with my video recorder are very much part of the story.)

This is not to say it was all an act. I did record the songs, the students were exposed to their elders' practices. But what was news was not the songs themselves but the inscription of tradition. The news, what was noteworthy, was the productive practices of salvaging and recording. The practices themselves lent urgency to the issue. The instruments inscribing culture themselves became objects of interest—as politicizing signifiers and exemplifications of endangerment in their own right.

Salvaging Nostalgia

Lantau Island is overgrown with significances. For many urban dwellers, Lantau epitomizes Hong Kong's past; it is mostly undeveloped, lush and green, a place where people live in old villages, where cars and high-rises are scarce, where you can breathe clean air, where people eat vegetables that they might have grown themselves. *This is what Hong Kong used to be like*, people whisper or pronounce to one another, as they walk through Lantau's green space, point at village buildings, and look at its residents.

Yet even while conjuring Hong Kong's past, Lantau also holds the promise of Hong Kong's future. Several times larger than Hong Kong proper, Lantau offers to planners and developers a relatively clean slate and space to accommodate Hong Kong's burgeoning population. Lantau is a blank page in a crammed notebook. It is the stuff of which dreams of Hong Kong are made.

What have these dreams consisted of? Already, Lantau Island houses Hong Kong's mega air terminal on its north shore, two new suspension bridges linking the airport to the peninsular New Territories, a small town redeveloped into an exurb over the course of the airport's construction, and an exclusive bedroom community on the southern coast occupied by expatriate and local professionals seeking refuge from the densities of urban life. In 2005 a bay on Lantau's eastern side would come to accommodate Asia's newest Disneyland. In the wake of a regional economic slump, government planners have pinned their hopes on a bright touristic future for the island. There are all the temples, they enthuse, the hiking trails, the largest outdoor seated bronze Buddha in the world at the Ngong Ping Monastery, where you can eat a Buddhist vegetarian meal. And, of course, there is Tai O.

If notions of Hong Kong's past and future weigh heavily on Lantau Island as a whole, they are perhaps nowhere more condensed than in Tai O. The village serves Hong Kong as a kind of cultural repository. Its name conjures images of salted fish and seafood, quaint old women in wide-brimmed straw hats speaking unintelligible dialect, and, inevitably, Tai O's celebrated paang uk (stilt houses; literally "stage houses")—houses built over the water. The homes are living spaces built from scraps of wood salvaged from boats, tar paper, sheet metal, houses supported on timbers sunk deep into the coastal mud. Tai O recalls for almost everyone in Hong Kong a visit to the village made with family or friends, reminds them of a rope-drawn sampan that crossed the canal between the bus stop and Tai O proper, the smells of shrimp paste, and the sight of salted fish hung up to dry or arranged in tidy spirals and starbursts on flat woven baskets.

Tai O is growing old. With the decline of the fishing, salt, and salted fish industries that anchored the economy through most of the last century, most Tai O people of working age have had to move away in search of employment. They have gone to Kowloon, Hong Kong, or the New Territories, leaving behind a village inhabited mostly by elderly people and school-age children. "This is a louhyahn kui!" chuckled one elderly woman to me one day while she and her friends played cards, "an old folks' neighborhood."

Tai O's age has not hurt its touristic appeal. On weekends the village streets, at least those near the bus stop and pier, are filled with visitors. Tourists from Hong Kong and abroad have for many years now come to Tai O to wander its web of narrow alleys, canals, and apparently aimless bridges, to watch old women and men knitting nets, and to eat seafood cooked at one of Tai O's restaurants after buying it from the man on the street who sells fish, clams, and crabs from blue rubber tubs of gurgling water. "Did you catch this, grandpa?" they ask. Here, a family in fleece; there, a young, slender couple in bell-bottoms and sunglasses; there a group of photography lovers in khaki vests snapping shots of stilt houses, drying fish, old people. The visitors almost always stop at the corner of Wing On Street, at the bridge, where the older members of each party will tell the younger ones that, in the past, there was no bridge here. No, before, you used to pay fifty cents to an old popo, or *sam*, a granny who towed you across the inlet on a rope-pulled boat. "Granny, is this where the pull boat was?" a braver one will ask the bemused woman sitting behind her boxes of eggplants, melons, and greens.

On weekdays, the streets are calmer. Many of Tai O's silver-haired residents rise at five o'clock in the morning and head to the Venice Café (named after the moniker given to Tai O by the Hong Kong Tourist Association— "The Venice of Asia") for a bite of dim sum and a pot of tea before taking a seat somewhere. Some might sit on a bench under a large tree or at some tables and chairs assembled near the main road, while others may gather at the foot of the Tai O bridge—all places with a nice breeze. Around 11 o'clock, the schoolchildren empty the classrooms to go home for lunch; their elders call greetings to them as they run, bike, and walk by in their uniforms. Some of the frailer villagers go back to their stilt homes or apartments to await a lunch box delivery from Senior Services. After lunch, many of the *louhyahn ga*, or old folks, lie down for a short nap. Others meet for conversation or card games. Perhaps they go to the YWCA Community Center, where Rosanna is holding a meeting with residents who want to urge the government to install a new street lamp on one of Tai O's walkways. Occasionally, a German tour guide with slick, blond hair leads an English-speaking tour through Tai O's main streets, stopping to pose near a shark fin hung in one of the storefronts on Wing On Street. By and large, though, Tai O receives the bulk of its visitors on weekends.

Tai O, for its history as one of Hong Kong's busiest ports, for its charm as a place of old fishing people and traditions, and for its prospects as an even greater magnet for Hong Kong tourism, sits in the midst of intense temporal meanings.

• • •

What do we make of the nostalgia that seems to inhere in the politics of endangerment? For what is endangerment if not an anticipatory nostalgia? It figures a lonely Old, threatened by the New. The threat to Tai O's *present* is glossed as a potential loss of Hong Kong's *past*. Endangerment invokes a need to protect Hong Kong's present from the future. The past is to be protected from the present, while the present is to be protected from the future; both are to be sheltered from the movement of history. This is a nostalgia that is almost impossible to wash oneself clean of. It saturates the senses—

the smell of shrimp paste, the sight of the sky and beaten-down buildings, the sound of quiet.

I invoke a supersaturated experience to suggest how visceral and immediate this nostalgia feels, how apparent its grounds seem to be for many people in Hong Kong, but I should be clear about two things. First, the sense of this place as nostalgic is not a given—its experience as precious past and old is itself relatively new. Second, the olfactory, aural, and other sensory cues I offer here are not meant to imply a prediscursive or precultural ontology of bodily sense. I wish to convey instead how central the development of capacities and tendencies to sense Tai O in particular ways is to the politics of endangerment, as well as to other ecologies and politics considered in later chapters.[12]

Not long ago, for instance, visitors to Tai O would have found themselves in squalor. I surveyed friends and relatives in Hong Kong, Canada, and the United States who had grown up in Hong Kong in the 1950s and 1960s. Had Tai O always been such an attraction? I asked. Did they have fond childhood memories of the place? Most people told me they did not. Some remembered hiking to a monastery nearby; but the Tai O of their memory was a dirty place with squatter huts, not a destination in its own right. Today, though I cannot pinpoint when it began exactly, Tai O has come to mean something else. Tai O used to be simply dirty and backward. Now, it stands for Hong Kong's history.

This is allochronism in raw form. Here, Hong Kong's airport and the vision of Lantau as a leisure island mark Hong Kong's new future. Dolphins and fishing villages, whether one is worried about saving them or resigned to moving beyond them, become the old and bygone. Alongside and within the epistemic work of the politics of endangerment, then, bodies caught in the field of Tai O's endangerment become sensate in a very particular way. Or, better put, part of the politics of endangerment is the emergence and cultivation of certain ways of experiencing the environs—ways of smelling, walking, listening, recording, photographing, remembering. The occasional barrages of sensory detail I present in this book mean to condense and amplify this emergence and cultivation. This is what it feels like to fear that one of Hong Kong's dahksik, a precious piece of its history, will soon be lost.

The preoccupations with old Hong Kong that manifest in the busloads of tourists exploring Tai O's narrow streets, snapping photos of stilt houses, do not tell a new or isolated story. Marilyn Ivy has described a similar phenomenon in Japan, where urban residents began in the 1970s to scour the Japanese countryside in search of traces of an authentic Japan whose vanishing they could mourn.[13] This search for, production of, and mourning of Japan's vanishing past, Ivy suggests, constitute the very meaning of "modernity" in Japan. Similarly, the positing of Tai O as Hong Kong's past marks ever so clearly how far Hong Kong has come, how much the vaunted financial hub has changed.

Recognizing the nostalgia and allochronism that runs through Wong Wai King's and others' experiences of Tai O can be discomfiting. Two decades ago, the anthropologist Renato Rosaldo provocatively dubbed "imperialist nostalgia" the mode or affect that he noticed characterizing the work of Western anthropologists and critics who described and bemoaned cultural loss or natural degradation in the global South.[14] What was especially pernicious about this nostalgia, Rosaldo helped make clear, was that through it we might mourn the loss of something while disavowing our own complicity in destroying it. It was imperialist nostalgia, for instance, when colonists lamented the transformations of native lifestyles in colonies. It was equally so when settlers decried the ravaging of a nature that they had helped to decimate, and no less so when tourists long for more "authentic" encounters with nature or culture. Thanks to Rosaldo and others, it is impossible to see such longing at work without asking after its simultaneously imperialist and romantic overtones.

The cultural studies scholar Eric Ma offers a salient criticism of nostalgia in Hong Kong that I find especially compelling. Ma focuses on the ideological work effected by the figure of the fisherman as icon of its history, particularly as it manifested in a highly acclaimed television commercial for Hong Kong and Shanghai Bank that aired in 1995, just prior to Hong Kong's shift in sovereignty. Shot in black and white and featuring a first-person voice-over, the commercial offers a fisherman's memories of surviving natural hardships—drought, rainstorm, typhoon—as a personalized history of Hong Kong. Ma argues that in addition to offering allegorical "reassurance in the face of the sociopolitical changes and discontinuity of the 1990s," the commercial presents "discourses of upward mobility and individualism in nostalgic disguise," encouraging an ideological and sentimental identification that naturalizes an entrepreneurial subject while obscuring the existence of a vibrant trade culture that existed before colonialism.[15] The nostalgic and self-congratulatory turn to the past, actualized through a documentary aesthetic and the figure of the fisherman, affirms the history that rendered Hong Kong a colonial and capitalist entrepôt, proffering a sense of local history only to navigate it "to a safe ground where global capitalism can coexist with a resinicized nationalism." [16]

Although I think Ma is right, I would not dismiss the nostalgia in Tai O too quickly. As anthropologist Kathleen Stewart puts it, the work of nostalgia "depends on where you stand in the landscape of the present." [17] It matters whether the nostalgic subject is a colonial officer, an anthropologist, a dolphin, a missionary, a Hong Kong tourist, a bank, or a village resident. It matters whether the nostalgic subject identifies — or is identified — with the new or the old.

Wong Wai King is a nostalgic woman. She has told me many times how much she loves Tai O and how much she appreciates old things. She is full of stories about how things used to be, and memories of being a young girl in the village. But her nostalgia, like the nostalgia of the women workers the anthropologist Lisa Rofel writes about, pries history open rather than suturing it closed. [18] Her actions fly in the face of the dominant sense of *waaihgauh*, a Cantonese word close in meaning to "nostalgia" that translates as something like "holding the old" (as one holds a child). It is to hold in one's heart what is gone. It does not necessarily mean that one wishes it never left.

Wong Wai King's projects, in contrast to this sense of waaihgauh, assert that the old is more than something to be remembered fondly; it can be something worth struggling for. They speak against the inevitability and acceptance of Hong Kong's development. This is a politicized nostalgia that Wong Wai King seeks to mobilize — even when it seems as futile as moving mountains.

Yu Gung and the Mountain

In 1993 Wong Wai King mounted a campaign to restore the seawall that used to keep the ocean off the salt beds that were central to Tai O's economy in the earlier half of the century. As the salt industry declined in Tai O, the seawall fell into disrepair. The wall's gradual disintegration deeply upset Wong Wai King, who harbored happy memories of walking along it as a child, of sitting on it and watching the sun paint the sky purple and orange as it set over the water.

Wong Wai King decided to do something about it. She approached the Rural Committee, the local governance office, where she was told that the seawall was of low priority; the more important thing was to build a larger harbor for the village so that boats could dock. The Planning Department didn't care either. Not to be deterred, Wong Wai King started asking schools in Hong Kong and Kowloon for help and notifying local newspapers of her intent. She named her campaign Yu Gung Moves the Mountain. "Do you know that story?" she asked me.

• • •

There once was a wise man named Yu Gung. Yu Gung had a beautiful home, a wonderful family, and he was very happy. Everything was perfect in his life, except for one thing. Two large mountains impeded passage to and from his home.

One day he resolved to do something about it. He and his sons went to the base of the mountain and began to dig. At the end of the day, they had moved a small mound of dirt and stones to the sea, which was a great distance from their home. Each day they went out and shoveled a little more.

His friends watched, amused, until one day they could no longer restrain themselves. "Yu Gung, you fool! You will never in your lifetime be able to move a whole mountain to the sea!"

Yu Gung looked up from his labor. "Yes, that is true. And my sons will never see that mountain moved either. But if my sons shovel, and their sons shovel, and their sons shovel, and their sons and grandsons after that, the mountain will eventually move."

• • •

"That was the idea I had in my mind for repairing the seawall," Wong Wai King recalls. "I knew I couldn't do it myself. But maybe with enough people, all working together, we could do it. And I knew that if I used the name of such a meaningful story, it could attract their interest."

It did attract interest. Over a hundred students from schools in Hong Kong, Kowloon, and the New Territories came for the special Yu Gung workday. Groups of volunteers hauled wheelbarrow after wheelbarrow of stones to the old seawall, where an old fisherman coordinated its reassembly. The press also covered it. And after a day of work, the seawall looked significantly better.

Fire

Wong Wai King succeeded in mobilizing university students, high school students, activists, reporters, and other concerned citizens in Hong Kong; but she had noticeably less luck with her fellow Tai O residents. "I would ask if they would come," she told me, "and they might agree with what I was doing. But they wouldn't want to help." She reasoned that this might be because they were elderly; perhaps they weren't as fit as she.

While most people I spoke with in Tai O thought that the government's plan to demolish the village's stilt houses made no sense, few people in Tai O were as exercised as Wong Wai King. Her friend Wong Chi Chuen, the YWCA social workers, and other activists and intellectuals from urban Hong Kong all made efforts to help, but most villagers refrained from entering the fray. Even if they professed respect for Wong Wai King, expressed pride in the book she had written about Tai O's history, and appreciated the attention her efforts brought to their village, they left her alone in her politics.

Why? I suspect it had something to do with the nostalgic framing of Wong Wai King's political projects. Though it powerfully attracted many people, particularly those living in more urban parts of Hong Kong, the temporal modality of her nostalgia, which emphasized the history of Tai O and posited the village as an endangered piece of Hong Kong's past, was not one through which many people in Tai O understood themselves. There were few places in Wong Wai King's nostalgic projects where her fellow Tai O residents would recognize themselves. And so they chose not to join her projects and politics.

Then, on July 3, 1999, everything changed. Somewhere in the tangled cables bringing electricity to the stilt houses by Tai O Creek, an electrical spark found a dry piece of wood, cloth, or paper, igniting an inferno that burned one hundred houses to the ground in the space of a night. Fortunately, there were no serious injuries. It had been a holiday and many of the young people who had come home for the weekend were able to help their older relatives get out of harm's way. Curiously, the fire burned down the exact set of houses the government had set out to demolish in its initial plan before the public outcry. Even more curiously to me, I heard no suspicions of conspiracy from anyone in Tai O.

I did, however, hear a great deal of anger. At a gathering held the day after the fire, a man ranted against the incompetence of the fire department. The firefighters, he said, had been unable to navigate Tai O's narrow, winding streets; they had taken twenty minutes just to find the right place. Once they arrived, they realized that they had brought the wrong gauge fire hose. To make matters worse, Tai O residents were prohibited from firefighting themselves. In the last fire, said the man standing next to me, Tai O people had saved their homes by pouring buckets of water on the roofs from the safety of boats.

Interestingly, the complaints echoed some of the very reasons that the Planning Department had used to justify their revitalization efforts. For planners, the stilt houses were a logistical nightmare: wiring was ad hoc and occasionally exposed; the walkways and bridges were narrow and crooked; there was no systematic infrastructure. The Tai O townspeople agreed with these statements, but they altered the nature of their consequences. These characteristics of stilt houses did not mean that they should be taken down or rationalized to make life easier for the government. No, it meant that the government needed to spend more time getting to know the roads.

A few days later, some men called a village-wide meeting. So many people attended that latecomers had to listen and watch from outside through the open windows.

• • •

Everyone knows that the point of the meeting is to come up with a way to demand of the government the right to *yundeichunggin*, to rebuild the stilt houses on the original plots. But first, we are told, we need to form an organization. What will we name it? One man suggests "Tai O Original Inhabitants Fire Recovery Committee." It's a bold move. The phrase *yungeuiman*, "original inhabitant," not only resonates with the demand to rebuild burnt homes on *yundei* (literally, "original land"); as the Cantonese term for "indigenous person," it also challenges the governmental classification of Tai O's stilt houses as squatter settlements. Though the fishing village existed prior to British colonization, it was not designated indigenous because the colonial authorities recognized only land-based villages.[19] Someone else points out, though, that many Tai O residents moved to the village in recent decades, and that the organization should include everyone who lives in Tai O. The room murmurs agreement, and a man in the back suggests "Tai O Residents' Fire Recovery Committee." That seems like a good option, but Chiu Kam Cheung, one of the men who had organized the meeting, interjects: That is a good name, but we need to think of this committee as being something that could talk to the government about what Tai O needs in the future too, even after this. The fire and the issue of rebuilding the stilt houses are certainly the most important thing at the moment. But they would still have many things to ask of the government. The town hall is full of nods of agreement, and with a unanimous show of hands, the Tai O Residents' Rights Committee is born.

. . .

The Tai O Residents' Rights Committee was very effective. It organized a march of three hundred people on the Hong Kong government building. It arranged meetings with the Planning Department and secured assurances from key officials that the interests of Tai O people would be of top consideration. By the time I left Hong Kong to return to the United States, the committee had filed a lawsuit against the government for reclassification and recognition as an indigenous village—on the grounds that Tai O predated colonial occupation—so that they might avail themselves of the limited land and building rights that attend that governmental designation.

While Wong Wai King had enjoyed success in mobilizing allies outside of Tai O through her nostalgic projects, the fire recast the issue of the stilt homes—tearing it from the idiom of cultural heritage in which Wong Wai King framed her work and inserting it into a language of loss and home. Though the Tai O Residents' Rights Committee would also declare that stilt houses were a Tai O dahksik, their focus on *yundeichunggin*, the rebuilding of homes on original land, downplayed the issue of historic value, emphasizing instead the problem of home loss.

Loss

The trees whisper as I sit in the shade near the bridge. Several Tai O residents are also enjoying the shade and breeze, grateful for a little distraction after the mayhem of Monday's fire, the protests, the press. The women, sitting on the metal benches under the trees, talk in hushed tones. "Poor Wong Chi Chuen! He lost the studio in his paang and all his paintings."

A group of men with firm round bellies stand at the edge of the bridge, griping in loud voices that make it sound as if they're angry. They swear and end their boasts and shouts with an emphatic *ah!* "That's what I said, it's not enough!" They talk about the cost of rebuilding a paang, how unaffordable it would be.

Next to me on my bench, Wai Popo quietly raises spoonfuls of black bean spareribs and rice to her mouth from a Styrofoam box, taking occasional sips from the can of orange Fanta by her side. I notice how small she is; her feet dangle half a foot above the ground as she eats.

Chau Popo shuffles by, leaning on her umbrella. "Eating?"

"Yes." Another mouthful.

"What kind of dish?

"Spareribs."

"Ah, spareribs." She peers at the tree for a while. "Where'd you get them?"

Wai Popo tells her she got the dish at the restaurant down the street as Rosanna, one of the social workers at the YWCA, walks by on her way to lunch. It's a hot, sunny summer day, but Rosanna seems unaffected by the heat and looks jaunty in a straw cowboy hat and plastic Snoopy sandals. Two popos sitting on the next bench over brighten noticeably and call out to her as she sashays by. "Hey, Lai Guneung! Miss Lai! What are you doing? Have you eaten yet?" No, she hasn't. She's on her way to get lunch now. Oh, you'll have to treat us sometime. Yes, yes, they all smile. Bye-bye. Bye-bye.

Then Mr. Wong, the broad-shouldered man who presented Tai O's list of demands to the police officer at Sunday's demonstration at the government building, approaches our little group. With him walks a man I haven't seen before. When Mr. Wong sees Wai Popo, still working on her mouthful of lunch, he nods and walks toward our bench.

"Wai Popo? This man is from Social Services. He wants to talk to you."

The man from Social Services kneels down near the bench. "Wai Popo, we heard that your paang burned down."

She nods.

"Your son's very worried about you . . ." he continues gently, "and he wants you . . . he wants you to move into the louyahn yun, the old folks' home."

Wai Popo turns to him with a surprised look, puts her spoon down. "What—where is it?" she manages.

"Tung Chung."

Tung Chung is a new town on Lantau's north shore. To call it a new town, though, is not entirely accurate, because until recently Tung Chung was a largely rural village. When the Airport Authority constructed the new Chek Lap Kok airport on reclaimed land north of Lantau Island, however, they also redeveloped nearby Tung Chung to serve as a town for airport employees. Tung Chung now boasts its own top-of-the-line Mass Transit Railway

(MTR) terminal, several public and private high-rise apartment buildings, and a shopping center with a Park and Shop superstore and a Häagen-Dazs ice-cream shop. A trip from Tung Chung to Central, Hong Kong's downtown area, takes only twenty minutes via a new MTR line, but the winding bus journey from Tai O to Tung Chung still takes close to an hour.

"I want to live in Tai O," the old woman says softly.

"There's no room in Tai O, Wai Popo. But there's room now in Tung Chung."

"It's so far!"

Mr. Wong jumps in helpfully, "Someone can bring you back to Tai O for visits."

"But I want to live in Tai O."

"It'll just be temporary in Tung Chung. There's a waiting list in Tai O."

"How's your health, Wai Popo?"

"There's no room in the Tai O old folks' home."

They go on like this for several minutes, until either Mr. Wong or the man from Social Services, I don't remember which, blurts out, "You have no *deifong*, you have no place. What other choice do you have?"

With that, the conversation ends. Wai Popo reluctantly agrees to move. "One day you're raw, the next day you're cooked," she says sadly.

Mr. Wong looks satisfied; he has found housing for an elder in need. He ushers the man from Social Services somewhere else, leaving Wai Popo to sit with her decision and cold spareribs.

"I want to live in Tai O," she repeats. She says it with feeling.

Chau Popo, like me, has witnessed the whole conversation from close by. She fidgets uncomfortably with her umbrella as the import of the conversation sinks into Wai Popo's face. "Tung Chung has lots of Tai O people," she offers awkwardly. "It'll be the same."

"I don't know anyone in Tung Chung. I want to live in Tai O."

"Eat first, don't talk so much," she gruffly admonishes her friend. "Eat."

"I want to live in Tai O."

"There's no room, didn't you hear him?"

"I can't eat . . ."

"Eat la!"

"I want to live in Tai O."

"If they have room, they'll move you. For now, you have to eat."

"I don't know anyone . . ."

"Eat la, don't waste it la."

"I'm not hungry . . . I have no appetite." Wai Popo puts her box of food down on the ground, looks sadly to some distant place.

"You're not going to eat? At least don't put it on the ground, there are flies." Chau Popo bends down to retrieve Wai Popo's lunch and places it gently on the bench.

The leaves rustle for a while.

Then Chau Popo breaks the silence with a sigh. "*Mou gai*, ah. There's no way out. Live there for a while, and you'll meet people. You will. There'll be lots of old people . . . There's no need to be scared. Hey, remember the other day when we gambled with grandpa? That was fun. We gambled all night."

The leaves rustle some more, and the men swear some more about the cost of building paangs.

"There's breakfast there, and there's dessert. It'll be nice."

One of the spunky grandmas sitting on the next bench over spies Rosanna walking back to the YWCA office and chuckles in anticipation, "We'll have to peg Lai Guneung when she comes by!"

"Don't be afraid," continues Chau Popo. "What are you afraid of?"

Wai Popo's quiet now, lost in her thoughts.

Chau Popo must sense she's losing ground, for she now gestures toward me, drawing me into their conversation for the first time. "It'll be okay. Like this gentleman . . . he's from Hong Kong. He came to Tai O. We didn't know him before, but now we do. Now we're friends." She shifts to address me. "Right?"

I nod vigorously, "Yes, yes we are."

"See?" Chau Popo turns back to her shrinking friend. "It'll be okay."

Wai Popo is unconvinced. "I don't want to move," she whispers, almost to herself. She sips some Fanta through her straw. "Just thinking about it makes me want to cry."

• • •

Why limit our reading of nostalgia as a longing for the past? This temporal take on nostalgia is common, but there is no need to jump so quickly into a temporal frame. The English word "nostalgia," as countless commentators have observed, comes from the Greek, denoting a kind of painful homesickness. How odd that, today, we so often take this "home" for which nostalgia aches in purely temporal terms. A politics of endangerment might offer an alternative definition of nostalgia—or, better, a reminder of another sense that has been there all along—where the spatial aspects of a desired return home are retained. If we remember this sense of nostalgia, we might see that nostalgic discourses of endangerment do not simply bemoan the passage of time, but are sick, instead, from the loss of specific, meaningful spaces.

On Coevality

I have sought in this chapter to demonstrate and to argue that the politics of endangerment rests upon two fundamental gestures—those of threat and specificity. These twinned gestures, I suggest, are particularly resonant in Hong Kong today, as issues of cultural specificity and political autonomy continue to preoccupy pundits and politicians, and as questions of economic niche—particularly as China's entry into the World Trade Organization (WTO) erases Hong Kong's privileged position as gateway between China and the West—worry Hong Kong capitalists.

The politics of endangerment discussed so far worked through two ecologies of comparison, the first temporal, the second spatial. In the first, the comparisons cast the endangered subject in relation to an idyllic past and apocalyptic present or future. Here, endangerment functions as an anticipatory nostalgia; an expectation that something of the present will, in the near future, be lost. Like nostalgia, endangerment moves along an imagined timeline, retrieving something from earlier, recasting it as later. But within the conceptual coordinates of this temporal ecology, endangerment positions its subjects in the future, looking backward, watching with dismay at the ruining of our present. And because it proffers this clairvoyant view, it can engender politics—because with foresight, the future can be changed.

Although reckoning nostalgia in purely temporal terms is a politically fraught enterprise, I offer two rationales for salvaging nostalgia for a politics of endangerment. First, nostalgia can, in the hands of the relatively marginal, offer a potent critique of the present and a rebuttal to allochronism. Second, and perhaps more importantly, the ache of endangerment's anticipatory nostalgia illuminates a great deal when the loss it forecasts is remembered to be both spatial and temporal.

This effects a conceptual shift, one that moves endangered forms of life from a timeline on which they occupy the slots of the old and bygone to a field of simultaneous, coeval spaces. At the same time, the examples of Tai O and the dolphin point out how the nostalgia in Hong Kong's politics of endangerment can reframe the dominant stories about Hong Kong's past and future. This, I think, is the promise of environmental politics—that through them not only might the allochronic machineries of "development" and "modernization" be refused, but a new field of political coevality might be configured. Within the frame of an anticipatory nostalgia that scans anxiously across both time and space, Tai O and the dolphin are unique and worth saving—not because they are pieces of Hong Kong's past that should not be sacrificed for the future, but because they have the right to coexist in the present with urban Hong Kong.

Like all political forms, however, endangerment has many faces. The form of life entailed and required by the logic of endangerment, what I will call "specific life," just as easily supports political quietism. I will explain this in chapter 3, with some help from a small orchid, another of Hong Kong's dahksik.

"Tim Go," the old man called out to me as we rocked back and forth in the bus, "you only need to learn one thing about Tai O." Around us sat the rest of our group, a lively bunch of elderly Tai O residents out on a field trip organized by the Tai O YWCA Community Center. Our destination was the supermarket in Tung Chung, where a few of the women on board would soon haggle with bemused grocery clerks at the deli counter over the price of a soy sauce chicken. I had met Sai Kau, who had been a fisherman in his younger years, through Rosanna, the YWCA social worker who had organized the trip, and we had boarded the bus and headed to the rear together. Now, about twenty minutes into the winding drive around the top of Lantau Island, Sai Kau engaged me with a conspiratorial look, and I leaned in, intrigued.

"Everything," Sai Kau revealed with a satisfied smile, "everything happens slowly here."

I nodded and tried to hide my disappointment. I had heard variants of this before. People from urban and rural areas alike often said that the pace of life in Tai O was slower than in Hong Kong. Another man had told me that people even walked slower there. The common sense of things was that in Tai O life was more relaxed, more humane. It was not as *ganjeung*, not as stressful or excited. This was one of the many ways in which Tai O residents habitually, casually demarcated themselves from Hong Kong. I wasn't surprised to hear Sai Kau say the same thing. I just wished he had said something more interesting.

But Sai Kau wasn't finished. Everything was slow, he repeated. Like the time they wanted the bridge at San Kee fixed, he continued. They had pestered the government to do it, and the government kept saying they would. But they never did. Every time people in Tai O wanted to get something done, they had to wait forever. He had lost count of the number of letters he had written to the government, with Rosanna's help, asking them to fix the San Kee bridge. Each time he registered his complaint, Sai Kau was told that it would be taken care of shortly. Each time, nobody came. "Government says they want to *faatjin*, to develop, Lantau Island," Sai Kau continued, "but they don't do anything for us. The only thing they do all the time is this!"

With "this," he stabbed his finger toward the window, pointing to the concrete hillsides girding the road. Steep slopes in Hong Kong were routinely concreted over to prevent mudslides. James Fong, an otherwise mild-mannered air pollution modeler at the Environmental Protection Department, once told me that these paved mountainsides drove him to distraction. They maddened him to no end because nothing would grow beneath the concrete, of course, which meant that there would be no roots to hold the soil together. So the soil would shift, the concrete would crack, a slide would ensue. The mountain would be repaved. And nothing would grow, the soil would shift . . .

Now, Sai Kau pointed me to the hillsides again. "Look at these! Every year there's a crack. They take the whole thing down and do it again. They keep doing it. Put it up, take it apart. Put it up, take it apart. Over and over again. But in Tai O we can't get a bridge fixed."

SPECIFIC LIFE

It was just a thin stalk spiraled by small white flowers, not an especially im-
pressive plant, but the two hikers crouched over it were more attuned to
plants and their peculiarities than most walkers of Hong Kong's countryside
in the 1970s. The taller of the two was Gloria Barretto, a botanist at Hong
Kong's Kadoorie Farms and Botanical Garden who had recently established
an orchid haven at the Garden. Her companion, a slight woman with large
glasses, was Hu Shiu-ying, a Harvard-trained botanist working at Harvard's
Arnold Arboretum. Hu was preparing to publish what would become the authoritative text on Hong Kong orchids, *The Genera of Orchidaceae in Hong Kong*,
and she frequently visited the Chinese University of Hong Kong while conducting her research. She and Barretto were fond of hiking together during
Hu's stays. I imagine that they were good company for each other, these
two longtime lovers of Hong Kong landscape and flowers, walking, talking,
enjoying the warm and relatively dry spring, their pulses quickening at the
prospect of sighting undocumented orchid species.

The flower the two botanists now saw—perhaps near the Chinese University or Hong Kong University campus, two of their favorite places to hike—
interested them. Hu and Barretto had been observing flowers like it for nearly
seven years. It was an orchid, in the genus *Spiranthes*, and its spiraling form
resembled that of another plant they knew well, *Spiranthes sinensis*. Unlike *S.
sinensis*, however, whose flowers were normally pink, this orchid had creamy
blossoms; and its bracts, sepals, and ovaries were covered with fine hairs,
unlike the smooth glands of *S. sinensis*. The two women had also noticed that

while this plant grew in sunny open landscapes similar to those frequented by S. *sinensis*, it preferred localities that were "damp but well drained."[1]

They logged their observation of this flower, and soon Hu and Barretto staged a more public appearance for it in a 1975 article, "New Species and Varieties of Orchidaceae in Hong Kong." In that article, Hu and Barretto claimed both distinction and similarity between this "new" orchid and S. *sinensis*. They named their new orchid *Spiranthes hongkongensis*.

Hu and Barretto's article identifies not only S. *hongkongensis*, but also twelve other undocumented orchid species. The bulk of these Hu and Barretto named after friends and colleagues, mostly residing in Hong Kong, who also appreciated, collected, and grew native orchids. There is, for instance, *Goodyera youngsayei*, named after J. L. Youngsaye, who had in the 1930s "amassed a considerable living collection and took excellent photographs" of orchids in Hong Kong; there is *Cheirostylus jamesleungii*, named after James Leung, "who discovered and collected the plant in flower"; there are *Liparis ruybarrettoi* and *Ania ruybarrettoi*, two flowers named after Gloria's son, Ruy, a lawyer hailing from one of the most prominent Portuguese families in Hong Kong and a board member of Kadoorie Farms and Botanic Garden; and there is *Gastrochilus holttumianus*, named in honor of "the foremost orchidologist for southeastern Asia," R. E. Holttum, "former director of the Botanic Gardens, Singapore, now at Kew."[2]

These names, marrying tropes of discovery and ownership, remind us just how wound up in one another are taxonomies, names, and figurative practices of exploration, discovery, collection, and property. Even so, beneath their explorer surface lie expressions of intimacy and expert care. One orchid, *Malaxi parvissimia*, we read, was "first collected and transplanted by W. G. L. Allan and has flowered repeatedly." Allan's success is then replicated. "According to [Allan's] direction, James Leung found another plant which was transplanted to the Barretto garden and also flowered repeatedly."[3] Amid rhetorics of measurement and technical description we can glimpse gardeners, friendships cultivated on hikes, and careful skills with plants that are notoriously difficult to make flower. Barretto apparently had quite a touch: remarks about repeated flowering in her garden appear throughout Hu and Barretto's report, bolstering a textual authority based not only on technical knowledge but on care and a flair for plants. Vegetable matter and intellectual ownership are folded into both civilizing Latinate names and a poetics of productive care, cultivation, and blooming.

In their report, Hu and Barretto name S. *hongkongensis* and another or-

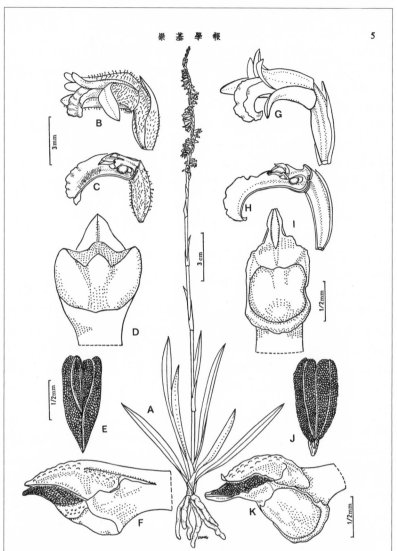

Figure 2. *Spiranthes:* A. The habit sketch of a flowering plant of *S. sinensis*, showing a fascicle of fleshy roots, linear basal leaves, and a flowering scape with spirally arranged flowers in a spike. B-F. *Spiranthes hongkongensis:* B. The lateral view of a flower with glandular hairs on the ovary and sepals. C. The lateral view of a flower with dorsal sepal, lateral sepals, 1 petal, and one-half of the lip removed, showing a spherical callus on one side at the base and the hairs on the disc of the lip, the column with an anther above and the stigma beneath. D. The front view of an apical portion of the column showing the middle of two pollinia, a very narrow hyaline rostellum attached to the middle of the pollinia, and the shield-like stigma trilobed along the upper margin. E. Four pollinia in 2 pairs, without evident viscidium. F. The lateral view of the apical portion of the column showing the pollinia, and the stigma attached to the middle of the 2 larger pollinia. G-K. *Spiranthes sinensis:* G. The lateral view of a flower showing the glabrous ovary and sepals. H. The lateral view of a flower with dorsal sepal, lateral sepals, 1 petal, and one-half of the lip removed, showing a spherical callus at the base and no hair on the disc of the lip, and the column with anther cap, pollinia, and stigma. I. The front view of the apical portion of the column showing the forked rostellum, the elliptic viscidium, and the oblong stigma. J. Four pollinia in 2 pairs attached to distinct elliptic viscidium. K. The lateral view of the apical portion of the column, showing the anther deep into the clinandrium, the rostellum bifurcate at apex, and the stigma.

Figure 1. Morphological Difference. (Hu and Barretto, "New Species and Varieties of Orchidaceae in Hong Kong," 5)

chid, *Manniella hongkongensis*, differently than they name all the others. These two names tie their plants not to a particular person but to a geography. Of course, there is an explicit trace of a natural history of colonialism in these names: the taming and naming of places in the heyday of European colonialism occurred in tandem with the naming of plants, animals, and landmarks, and the natural history of orchids in Hong Kong belongs to this history. From the 1870s and 1880s onward, British residents of Hong Kong collected orchids, painted pictures of them, and wrote about them; and they sent their specimens, paintings, and manuscripts to London and Kew, where plants not yet described by Western natural historians received specifying, or speciating, names. It was in this early colonial period that we see some of the first *hongkongensis* appellations—*Ania hongkongensis, Thelaris hongkongensis, Tropidia hongkongensis*—and it is impossible from today's vantage point to separate these proud new regionalizing plant names from the proud acts of region-naming that characterized the British Empire.

To mention only the colonial legacy in the orchid's naming, however, would miss something crucial. Another, immediate rationale for the name lies in the material practices that authorized any new name at all and the ecology of comparison within which Hu and Barretto were working in the 1970s. The *hongkongensis* name holds within itself an argument in miniature, an argument of distinction, a claim that it is not the same as *S. sinensis*, an orchid that (unwittingly) sets the terms of the debate by itself having a regionalizing name. Look at the dense exchanges of report, consultation, and comparison in the botanists' account of their justifications for the claim of specificity and speciation. They write that after seven years of careful study of the *Spiranthes* species, they have discovered that there are in fact two types in Hong Kong, one with smooth glands and the other with hairy ones. They nod to someone named Loureiro who was the first to write about *Spiranthes* while stationed in Macau and Guangzhou, describing in 1790 a flower that later came to be known as *S. sinensis*. Hu and Barretto then invoke an ally in Paris, Professor J. F. Leroy, with whose help they are "able to ascertain that Loureiro's material represents the glabrous element."[4] In other words, they propose that the flower described by Loureiro accounts for only one of the two kinds of flowers (the smooth one) that have typically been called *S. sinensis*.

Aided by the collaborative comparisons of their network of orchid appreciators, Hu and Barretto are poised to make their case for adding a new distinction—a new specificity/species—to orchid taxonomy. The name *S. sinen-*

sis should refer only to the glabrous orchids, they assert. "The hairy element," they contend, "requires a new name, which is proposed here as *Spiranthes hongkongensis* S.Y. Hu et Barretto." Some pointed comparisons follow. The flowers of *S. sinensis* last a week; those of *S. hongkongensis*, by contrast, "wither within three days."[5] The ovaries of the former remain slender. Those of the latter enlarge.

In the heat of this comparative argument, the regional appellative is almost an afterthought, a convention used without note of its sticky legacies, while the botanists focus on making the claim of *species novum*. Could anyone have known that *hongkongensis* might be laden with other meanings after another shift in sovereignty?

• • •

I learned of Hu and Barretto's discovery thirty years after the fact, in the office of Dr. Mei Sun, a population geneticist in Hong Kong University's Department of Botany and Zoology. A few years ago, Sun told me, she had had the honor of meeting Dr. Hu. Had I heard of her? Hu was one of the world's foremost authorities on Hong Kong orchids, just recently retired. Sun smiled with fond respect as she painted a picture for me of an older woman, energetic and deeply passionate and knowledgeable about orchids, then told me about Hu's collaboration with Barretto.

This flower that Hu and Barretto had discovered and named, Sun went on, had intrigued her. Was I familiar with electrophoresis? she asked. Sun, it turned out, was a master of this technique used in genetic tests to distinguish and measure relative amounts of various proteins and enzymes. She had earned a reputation for producing remarkably clear bands and separations in her gels, something that takes years of practice and a talent for fine-tuning instruments and solvents so that proteins identical in almost every way will be moved by an electrical current through the starch and gel mediums at measurably different rates by virtue of their small difference. She had a knack, in other words, for running gels that produced visuals that clearly distinguished one thing from another. Sun had become interested in both *S. hongkongensis* and *S. sinensis*, curious whether her expertise in electrophoresis might help her to determine whether the two flowers were indeed, genetically speaking, distinct species.

Sun's results were significant. The naming distinction that Hu and Barretto had adopted on the basis of morphological differences could be supported by evidence of genetic difference. S. *hongkongensis* was genetically different from S. *sinensis*, possessing a different frequency of certain alleles than the reference species. Furthermore, Sun had studied the reproductive and breeding patterns of both flowers. The two could cross under some conditions, producing a hybrid, but their offspring could not themselves reproduce; they were sterile. Hu and Barretto had been proven correct. S. *hongkongensis* was a genetically and reproductively distinct species in its own right.[6]

We are suddenly back with specificity and species, themes that preoccupied us in chapter 2. Specifying practices harmonize genetics with botany and natural history, linking orchids with Hong Kong's white dolphins, grouping flowers and marine mammals with the dahksik of fishing-village culture. Sun's work parallels in form the work of the SWIMS biologists in its acts of comparison and distinction.

And yet, just as S. *sinensis* and S. *hongkongensis* are not only similar but distinct, so too is Sun's work not simply similar to that of the SWIMS biologists; there is something different going on as well. Enabled by her expert ability to produce evidence of genetic and reproductive distinctiveness — something the dolphin scholars could not quite do — Sun could also advance a theory of how this unique and distinct form of life came to be.

Sun narrated this for me that day in her office as a puzzle of kinship and provenance, part science thriller, part family drama. S. *hongkongensis* was not the same species as S. *sinensis*; this proved Hu and Barretto right. But overshadowing the discovery of species distinctiveness was another finding: the two species were kin. She paused for effect. In fact, she said, S. *sinensis* was S. *hongkongensis*'s mother!

I was stunned. Here was a remarkable tale of plant kinship, parenthood, and specification, a tale of simultaneous relatedness, genetic indebtedness, and present incompatibility. It was also a tale that resonated surprisingly with questions raised by economists, political scientists, and other social scientists about the best way to understand the relationship between Hong Kong and the motherland after the handover in 1997. What was one to make of this resonance between plants and politics?

• • •

If only the distinctiveness of a cultural or political form of life could be substantiated as forcefully as the genetic autonomy of a flower! If they had thought to consider it, anthropologists and other cultural critics writing

	S. hongkongensis	Hybrid	S. sinensis	
Aat-2	aadd	add	dd cd cc bd	Aat-2
Aco-1				Aco-1
Aco-2	1bbbb 2bbcc	1bbb 1bbb 2bbc 2bbc	1bb 1ab 1aa 1bc 1cc 2ab 2ab 2bb 2bb 2bb	Aco-2
Est-1	bb	ab	aa	Est-1
Lap-2	bbcc	abc	bb bc ab aa	Lap-2
Mdh-1				Mdh-1
Mdh-2	1aa 1aacc 2bbbb 2bbbb	1acc 1acc (1ac)(1ac) 2bbb 2bbc	1cc 1bc 1bb 2bb 2bc 2cc 2ab 2aa 2ad 2ac 2cd	Mdh-2
Pgi-1	aabb	aab	aa ab bb	Pgi-1
Pgm-1	aacc	acc abc ace	bb bc cc ce ee ed dd de be	Pgm-1
Skdh-1	aa aabb	abb abc ad	bb bc cc cd dd bd	Skdh-1
Tpi-1				Tpi-1
Tpi-2	1bbbb 2aabb	1abb 2aab	1ab 1aa 1bb 2aa 2aa 2aa	Tpi-2

Fig. 2. Some of the observed zymograms at 12 polymorphic isozyme loci in populations of *S. hongkongensis, S. sinensis*, and their natural hybrids, showing gene duplications or "fixed heterozygosity" in *S. hongkongensis*, diploid banding patterns in *S. sinensis*, and the characteristic triploid phenotypes in the hybrids. The thickness of the band indicates the relative isozyme activity, reflecting the underlying dosage effect of gene duplication in the nonsymmetrical zymograms such as in the hybrids.

(*S. sinensis*) and *P2* (*S. spiralis*) were 0.884 and 0.895, respectively. The value of genetic identity between *P1* and *P2* was 0.60, higher than 0.487 obtained from the present samples of *S. sinensis* and *S. spiralis*. The difference between the two estimates may reflect genetic divergence between the parental materials involved in the

formation of *S. hongkongensis* and the samples used in this study.

Reproductive biology and breeding system—Both *S. sinensis* and *S. hongkongensis* flower in early spring in Hong Kong. However, *S. hongkongensis* starts flowering in the 1990s might have envied Sun's abilities to render difference visible. For they too were embroiled in acts of specification, though in an idiom of cultural rather than biological difference. Think of the argument made by Gordon Mathews, the anthropologist I mentioned in chapter 2 who reported that a new social category, *heunggongyahn*, had emerged in Hong Kong. Mathews remarks that Hong Kong identity is something always understood in relation to—yet distinct from—Chinese identity. For those who consider themselves *heunggongyahn*, he observes, the most important characteristics of a Hong Kong identity seem to be markers of affluence and consumption, the use of Cantonese and English as opposed to Mandarin, and a desire for democratic political ideals.[7]

And just as Sun followed the task of inscribing distinction (through her bands and separations) with a theory of speciation—a theory of how that difference, an orchid's specificity, came to be—so too did social and cultural analysts put forward explanatory accounts for the emergence of a Hong Kong culture distinct from that of China. For Mathews, as for many others, the speciation of *Homo hongkongensis* occurred because of a demographic shift that took place in the late 1960s and the 1970s, when for the first time, a generation of Hong Kong–born residents came into adulthood. There is remarkable agreement among students of Hong Kong society and culture about this period's significance. Prior to this time, the prevailing common sense goes, in the early colonial era Hong Kong was populated first by merchants and traders, then later by refugees from China fleeing the Cultural Revolution. Such people had no real ties to place, or if they did, it was to places of origin. The line distinguishing China from Hong Kong was remarkable mostly for its porosity. People, funds, remittances, and goods moved easily between mainland and colony. Different versions of this story of initial indistinction emphasize different things. Some might highlight the time spent by the revolutionary Sun Yat-Sen in Hong Kong; others focus on Chinese smuggling operations that brought rice and other staples to Hong Kong residents during the Japanese occupation. Still others might point to continued links of investment between Hong Kong and China. Whatever they emphasize, these narratives of original unity always serve as a background against which a Hong Kong self-identity gradually emerges. That emergence is always seen to begin with the demographic shift in the 1960s and 1970s and to be punctuated by Tiananmen.

The work of Maria Tam, another anthropologist writing on Hong Kong identity in the late 1990s, provides an exemplary case.[8] Tam is interested in *yumcha*, the practice of going out for dim sum. She argues that yumcha epitomizes the diversity and inclusiveness of Hong Kong culture, and that in the practice of yumcha, people reproduce physically, vividly, a desire for change (change from Chineseness) and an ethic of variety.

Tam's analysis hinges on its own tale of speciation: In the beginning, there were teahouses in Guangzhou where merchants might eat a few dumplings with a cup of tea. But in Hong Kong the practice gradually changed and expanded to the point where today people readily identify something distinctive as *gongsik yumcha*, or (Hong) Kong–style yumcha, which has great variety, where there are foods of non-Cantonese origin—including Western. For Tam this is a cultural artifact of Hong Kong's colonial history, with Southeast Asian foods signaling histories of exchange between Hong Kong, Singapore, Indonesia, Malaysia, and other countries in the region, and with gongsik yumcha itself exported beyond Hong Kong, to North America, Australia, and Europe as, ironically, a relatively fixed and enduring tradition for Hong Kong immigrant communities. Hong Kong yumcha thus derives from, but is significantly different from, southern Chinese dim sum, by virtue of its increased diversity, quantity, and eventually, delicacy. The pride that people express in gongsik yumcha and their fierce identification with it, Tam recognizes, enact a regional chauvinism enabled by the greater wealth enjoyed by Hong Kong relative to most of China. It hinges upon a certain fantasy of consumption, where capitalist economy yields greater consumer choice, greater abundance, and luxury. Tam's argument figures a food practice and identity predicated on certain ideas of cultural sophistication, grounded in an everyday cosmopolitanism that derives from Hong Kong's history as a capitalist entrepôt, a cosmopolitanism that reaches globally but that is also always localized. This localization of cosmopolitanism is precisely what characterizes Hong Kongness—a claim not far from that made by another anthropologist, James Watson, who writes that in Hong Kong, "the global is the local."[9]

Tam's argument appears simple at first. As I mentioned, it hews to the generic story of original unity and subsequent distinction. Tam supplements it, however, with something more nuanced than the typical argument about the emergence of a distinctive Hong Kong culture. She does not merely say that Hong Kong culture has emerged recently as a culture in its own right, but argues that what we might call Hong Kong metaculture—the very interest in and discourse on Hong Kong culture—has come into being as well.[10] Furthermore, Tam makes a provocative claim: because political autonomy for Hong Kong after the handover was known to be an unpopular topic for Beijing, local desires for autonomy were sublimated into the realm of culture. Assertions of cultural specificity and cultural identity were less threatening than struggles to attain locally elected government.

. . .

Could a sublimation like the one suggested by Tam—some substitution of biological autonomy for political sovereignty—have been occurring with Sun's revival of the recognition of a distinctive Hong Kong species? It is tempting for social scientists to frame scientific matters within the political. In the present case, this would entail invoking not only the handover but also the financial decline of the late 1990s and China's entry into the WTO— which in one stroke made Hong Kong and Shanghai rivals on equal footing as Chinese port cities accessible to world trade—to contextualize the broad range of endangerment tropes within the particular moment. To offer this analysis is to engage in a sociological reading, where sociological context is considered to be manifest in the form of the text or cultural object of concern. Here, the cultural object to be contextualized would be the recurring figure of biological and cultural specificity. The sociological reading—and it is at first blush a compelling one—would be that the historical period of political transition and economic questioning occasioned the widespread discourses of endangerment, of dahksik, of nostalgia.

This form of analysis has much to recommend it. It creates the impression that we have made sense of things. It anchors diverse phenomena in a particular situation, framing them against a common backdrop. We sense that we are documenting the wide-ranging implications of a world-historical moment of political crisis and transition. When I offer this form of argument to other social scientists, I know I will see nods of agreement and understanding.

Yet I find this approach unsatisfying for two reasons. First, such a reading is rigidly, and for no good reason, unidirectional; broader historical and political circumstances are figured as the cause to be read through the effect of cultural forms, even when the cultural forms include particular scientific and political practices. Reading in this unidirectional fashion underestimates the ways in which certain cultural forms come to structure our very sociological imaginations. Second, and this is a related point, it naturalizes the terms of sociological analysis. It proceeds as if Hong Kong *were* endangered and sees environmental politics as an expression of that sense of endangerment. This would mean the concepts for understanding economy and society are somehow less mediated or more solid than cultural objects, political formations, and their forms and logics. This is a problematic point of departure, both because expert pronouncements about governance and economy are no less produced through situated knowledge practices than expert claims made about forms of biological and cultural life,[11] and because it takes for granted a certain concept of endangerment, when, as we have seen, endangerment requires work to build.

What if we reverse the terms of the question? What might we see if we tried not only to read the terms and logics of ecology through sociology, but also to read sociology through ecology? Even if we pursue the argument that productions of Hong Kong cultural and political specificity expressed, or were symptomatic of, Hong Kong's historical condition, their recognizability as expressions of Hong Kong's situation should beg additional interpretive questions: What do we make of the fact that these knowledge practices seem such apt condensations of the problem of Hong Kong's sovereignty? What truth do these particular ecopolitical knowledge practices illuminate and make about Hong Kong politics?

. . .

The truth concerns the "nature" of autonomy. More to the point, it has to do with the conception of natural organic life that grounds the concept of freedom, which in turn grounds the very ideal of autonomy in Western political thought and, not coincidentally, in postcolonial struggles for autonomy.

This statement requires a quick philosophical detour to explain. Political philosopher Pheng Cheah, following the work of the historian of science Timothy Lenoir, has recently reminded political theorists that a particular figure of organic life structures the concept of freedom.[12] This can be traced through the work of Immanuel Kant, the Enlightenment philosopher whose discussions of human freedom became the ground for later writers' thinking about the value of political freedom and continue to be taught today in political science departments around the world as a foundation in political theory. The figure of the organism shaped not only Kant's thought, but subsequently the thinking of those—like Fichte, Hegel, and Marx—who sought to expand and concretize the ideal of freedom.

One can find analogies drawn between nature and society as far back as

Aristotle's comparisons of animal and human ontology, but the reliance of Kant's philosophy on comparisons with natural life is especially noteworthy because of a development in eighteenth-century German natural history and anthropology with which it coincided. The predominant view of nature had until then been of a machinelike totality, set into motion by an external and prior creator or principle. In 1781, however, a biologist and anthropologist named Johann Friedrich Blumenbach argued, with the aid of his microscope, that an organism could organize itself. He had observed that polyps, when cut up, would grow new parts. This, for Blumenbach, indicated that an organism was the cause of its own spontaneous action, in possession of what he called a *Bildungstrieb*, or formative drive.

Such a view of organic life, where an organism is vitally spontaneous, profoundly marked and inspired the thinking of Kant, who at the time was grappling with the question of whether and how humans could be free. He had, in his earlier *Critique of Pure Reason*, erected a stunning proof that human understandings were inescapably enabled by certain categories of thought (space and time) and faculties of perception. He viewed these categories and faculties as universally human.[13] Having established in the first *Critique* that there were a priori structures of human thought, Kant had then to broach a troubling question: Was it possible for humans to be free? If all human thought and experience was structured by the categories and faculties he had outlined, how could humans consider themselves the sources of their own actions?

We can imagine why Kant found inspiration in Blumenbach's theory of epigenesis, in which organismic life spontaneously organizes itself. By drawing analogies between human rationality and organic life, Kant could argue that an initial gift—of physical matter, of human categories—did not preclude the possibility of spontaneity. He looked for an analogue in human existence for the vitalism of the organism, finding it in the self-organization of society and the sphere of creative activities such as art, literature, and music. Cultural production produced novelty, and it could bring forth new organizations of human reason. This relatively untrammeled activity of *Bildung*, which referred both to cultural production and to its positive effects in molding and developing a human subject, is thus for Kant the analogue of organismic life and the paradigm for human freedom.

Pheng Cheah traces the profound effect these conceptions of organismic life and Bildung have had on political theory—from Kant's arguments that society's self-organization and culture are the analogues of organismic vitality, through Fichte's and Hegel's concretization of the ideal of Bildung in state form, through Marx's arguments that labor actualizes human creativity in the material world and that the proletarian class will self-organize, and up through socialist nationalist struggles for decolonization in Asia, Africa, and the Americas. For our consideration of the family resemblance between *S. hongkongensis* and a speciating Hong Kong culture and politics of sovereignty, the most relevant implication of Cheah's argument is that the concept of sovereignty that has characterized (postcolonial) nationalisms has arguably been structured and enabled by this Blumenbachian idea of organismic vitality and autonomy. The concept of nation, understood philosophically, was an instrument—conceived as a practice for transcending finitude, for surpassing the limits of mortal, natural, given life, and seen as a means to culturalize humankind. The struggle for autonomy is the struggle for life.

I find this genealogy provocative not only because it suggests that notions of political autonomy derive from philosophical efforts to conceptualize and practice freedom. Perhaps more important, it illustrates that the philosophical ontology of political freedom has been the natural ontology of organismic life. Struggles for autonomy are inseparable from concepts of freedom, which are in turn inseparable from theories about nature. Cheah shows that philosophies of freedom and subsequent concretizations of such philosophies in theories of political freedom derive from a particular understanding of the nature of natural life. Thus, anyone who has grappled with questions about the meaning and nature of life and freedom has grappled, knowingly or not, with what today would be considered the purview of the life sciences.

Keeping this in mind while thinking about Hong Kong's orchids, dolphins, and the questions of cultural and political autonomy raised at the end of the twentieth century enables us to see something remarkable. From the vantage of environmentalism and Hong Kong, the conceptual forms of organic life and political sovereignty look different than they did in national liberation struggles in the 1960s. Autonomy remains valued, but it has acquired a different valence. For the organism is no longer simply vital; it is ecological.

• • •

For Sun, the work of specifying *S. hongkongensis* genetically was inseparable from a warning of endangerment. "Hong Kong is home to many rare and endangered wild orchids," reads her project summary in an annual report of the Hong Kong Research Grants Council, "one of which is *Spiranthes hong-kongensis*." It continues, "A significant number of the orchid species now found in Hong Kong are endemic, suggesting a long history of orchid evolution and speciation in divergent habitats. However, conserving these rare species is becoming ever more challenging as Hong Kong becomes increasingly urbanised and the natural environment shrinks."[14]

Two things are striking about this brief passage. First is its invocation of endemism, what I describe in chapter 2 as the value ascribed to a form of life that is unique to place. Second is the proximity of the figure of endemism to notions of rarity and endangerment. Endemism is not linked to rarity or endangerment explicitly; the declaration that many orchids are endemic stands separate, but barely, instead tied to a statement about what endemism indicates about the length of history of evolution and the diversity of habitats in which speciation occurs. It is here that we can discern the work of a particular ecological logic, that of biodiversity. For from endemism and speciation—the two are inseparable in evolutionary ecology—a leap is made to rarity and conservation. "A significant number . . . are endemic, suggesting a long history of . . . speciation . . . However, conserving these rare species is becoming ever more challenging . . ." Endemism slides into an equivalence with rarity.[15] Specificity becomes a value in and of itself, and rarity is implied to be equivalent with a moral imperative for conservation.

What characterizes organic life here? No longer simply vital, as it was for Blumenbach and Kant, it is for one thing genomic. Sun's report, as well as numerous works in science and technology studies, makes this much clear.[16] Equally, though perhaps less charismatic than genomic life in its displacements of vital life,[17] is the conception of life as *ecological*—biological life with somewhat less freedom because of its embeddedness in systems, and highly specified through the informatting of species and genetic difference through population biology, genetic testing, and the building of biodiversity databases. Ecological life is also, significantly, a form of biological life that is unarguably political. The ecological sciences as we know them— including conservation biology and population biology—emerged in tandem with environmentalism. Environmentalists who struggled for years to articulate a rationale for preserving nature and a way to ascribe value to a form of life they wished to protect found a collaborator in genetic method. An environmental ethic paired with a conception of organic life as genetic code, paired in turn with the promissory aura of specificities unattached to name or patent—an aura reminiscent of, but different from, the furor to name flowers in the eighteenth and nineteenth centuries—yielded the modern notion of biodiversity.

One thing is clear: in the world of biodiverse life and endangerment, an organism's most salient characteristic is not its vitality. Auto-causality does not preoccupy scientists like Mei Sun, Hu Shiu-ying, and Gloria Barretto. Instead, what concerns them—whether in genetic or ecosystemic terms—is a form of life's specificity.

• • •

It is upon this conception of life—specific life—that Hong Kong's politics of sovereignty were built. A glimpse of this is offered in Ackbar Abbas's provocative book, *Hong Kong: Culture and the Politics of Disappearance*, published in the year of the handover. It begins with a tale of speciation similar to those of Mathews, Tam, and others who were writing on Hong Kong culture in the 1990s. In the beginning, there was no Hong Kong culture. But then, there was. One immediately notices, though, that Abbas's account takes pains to address itself not to the characteristics of culture, but to the characteristics of the discourse on culture: "One of the effects of colonialism was that until as late as the seventies, Hong Kong did not realize it could have a culture. The import mentality saw culture, like everything else, as that which came from elsewhere: from Chinese tradition, more legitimately located in mainland China and Taiwan, or from the West. As for Hong Kong, it was, in a favorite phrase, 'a cultural desert.' Not that there was nothing going on in cinema, architecture, and writing; it was just not recognized to be culture as such."[18]

Notice how Abbas works to ensure that his analysis will not be a part of the phenomenon he describes. His sentences address what "the import mentality" sees as culture, how "a favorite phrase" describes Hong Kong; Abbas is not describing Hong Kong, these turns of phrase announce, but rather how others see Hong Kong. In the beginning, because of colonialism, there is no (sense of) Hong Kong culture. It is not that in Hong Kong there were no creative moments in architecture, cinema, or writing. Rather, any developments in these arenas were interpreted through a lens that saw culture—the capacity for and realization of creativity—as originating anywhere but Hong Kong. Hong Kong culture is always bracketed, in explicit or rhetorical quotation marks.

Meanwhile, there is something Abbas has not quite named, something that would encompass this belief that Hong Kong culture does not exist, that will itself characterize Hong Kong. There may have been no "Hong Kong culture," but there is a "Hong Kong" in Abbas's formulation that does not realize its own culture.[19] This metaculture—which renders Hong Kong as having no culture—shifts in the late 1960s and early 1970s, the same period as the demographic shift that Mathews argued spawned a newfound category of identity. Yet whereas Mathews was keen to emphasize the deeply felt nature of this category and the ways such identification pointed to a new distinction between Chineseness and Hong Kongness, Abbas emphasizes the tie between concerns about Hong Kong specificity and what he characterizes as an anticipation of its imminent disappearance in 1997. This anticipation, Abbas argues, is what characterizes Hong Kong in the last days of empire.

> What changed the largely negative attitude to Hong Kong culture was not
> just Hong Kong's growing affluence; more important it was the double
> trauma of the signing of the Sino-British Joint Declaration of 1984 followed by the Tiananmen Massacre of 1989. These two events confirmed
> a lot of people's fears that *the Hong Kong way of life* with its mixture of
> colonialist and democratic trappings was in imminent danger of disappearing. "Anything about which one knows that one soon will not have
> it around becomes an image," Walter Benjamin wrote. The imminence of
> its disappearance, I would argue, was what precipitated an intense and
> unprecedented interest in Hong Kong culture. *The anticipated end of Hong
> Kong as people knew it was the beginning of a profound concern with its historical
> and cultural specificity.* But then the cause of this interest in Hong Kong culture—1997—may also cause its demise.[20]

Abbas's discernment of how a particular form of Hong Kong life was posited on the verge of disappearance helps to underscore the workings of the politics of endangerment. He astutely argues that the anticipation of 1997 was a cause for the emergence of Hong Kong specificity as an object of interest and concern. At the same time, perhaps intentionally, Abbas's language at times performs the very logics of endangerment and disappearance he seeks to illuminate. "The imminence of its disappearance" he writes: With his definite "the" and pronominalizing "its," Abbas presumes and ontologizes both "the Hong Kong way of life" and "its" coming disappearance in one fell swoop. The year 1997 comes to stand as self-evident cause—not only of concern but of demise. What Abbas calls "its demise," furthermore, entails more than one death, if we read through the figure of specific ecological life. "Its" is deceptively singular, for an entire people's specificity is at stake. Performed in this text is a threat of extinction.

This preoccupation with the future of a "Hong Kong way of life" after 1997, says Wing-sang Law, a professor of cultural studies at Hong Kong's Lingnan University, has become the generally accepted political correctness in Hong Kong and outside it. It is, without exception, the sentiment behind the first questions he is always asked when he lectures outside Hong Kong. Has life changed? Is there still a free press?

Law is concerned about this, asking pointedly, "Is Hong Kong politics about nothing but post-1997 freedom and democracy? Is concern for freedom and democracy adequate as a marker for cultural critics' political correctness?" Law takes issue with the way in which cultural workers focused almost exclusively on the issue of Hong Kong uniqueness in the late 1990s. Like Tam, he links the investment in Hong Kong culture to concerns about political sovereignty, saying that Hong Kong writers were "echoing the much politicized call for Hong Kong citizens to come out and protect their own 'ways of life'" when they "emerged from their secluded literary or artistic enclaves and demanded that people should care about Hong Kong culture."[21] But at what cost?

Focusing on uniqueness, and the supposed threat of that uniqueness's extinction posed by 1997, Law argues, has enabled a general eclipse of social critique within Hong Kong, a disavowal of other politics and other lives. In this way, he points out, cultural criticism disregards the politics of labor and social inequity, as well as the "northbound colonialism" of Hong Kong investment and influence in southern China.[22]

The importance of Law's critique can be clearly illustrated. Consider a statement made in 2002 by Norman Lyle, chairman of the British Chamber of Commerce in Hong Kong, about the importance of civil liberties. The statement was sent to Hong Kong's secretary for security, Regina Ip, in response to Article 23, an anti-subversion law proposed by Tung Chee-hwa's administration. Article 23 was roundly condemned by opponents, who mostly self-identified as pro-democracy, as an infringement on civil liberties. Lyle's letter was quoted in an American newspaper, the *San Jose Mercury News*.[23] Notice the political work toward which the ideas of specificity and freedom are leveraged in this excerpt: "'Hong Kong's strength lies in its respect for the rule of law and fundamental rights and freedoms,' Norman Lyle . . . wrote in a recent letter to Ip. Anything that erodes freedom, he said, 'will make Hong Kong a much less favorable location for international business to be based simply because it will remove its regional and national uniqueness.'"24

In tune with the discourse on cultural specificity, Lyle focuses on the question of Hong Kong's uniqueness. Freedom in this account is not spontaneous or vital life; it is not self-generated action. Freedom is instead one among several unique characteristics to be safeguarded. Much as Mathews names an appreciation of freedom as one in a family of characteristics that distinguished heunggongyahn from Chinese, the chairman makes freedom itself one of a number of characteristics that uniquely positions Hong Kong as a hub for international business. Democracy equals the preservation of that key uniqueness.

Organismic life—in articulation with Marxist and anti-imperialist movements in Southeast Asia, Africa, and Latin America—may have fed a politics of postcolonial liberation; but here, in this rendering of the need for democracy in Hong Kong, the articulation of endangered *specific life* turns the ideal of political autonomy into no more than a condition for the perdurance of Hong Kong business and tycoons. The ontology attending this particular ecological conception of political life thus differs fundamentally from the postcolonial project of actualizing freedom through the self-impelled nation. The project is not to transcend conditions of finitude. Instead, it is to continue living as you were.

A sobering thought. The pro-democracy mobilizations that emerged in anticipation of and in the wake of the transition to Chinese sovereignty were primarily conservative—they enacted a politics of conservation. Conservation and biodiversity as paradigmatic grounds for conceiving of Hong Kong democracy did not actualize life; they simply enabled efforts to preserve it, to protect the way things were. While many Western writers hailed calls for recognition of a uniquely Hong Kong form of life as a spontaneous call for sovereignty, the life at stake in the bid for autonomy was one whose fundamental nature was not its freedom, but its enduring, essential specificity.25 Autonomy was the means; persistence rather than vitality was the end.

• • •

Certainly, ecology is about more than endangered specificities. The ecological sciences teem with multiple frameworks for conceptualizing life, some competing, some operating in parallel. A different scaling of the problem designates a different unit for life's analysis and definition: a population, an ecosystem, a cascade of energy's transformations, a cycle of carbon, nitrogen, or other elements.[26]

I have purposely narrowed the field, focusing on the figure of specific life that emerged through the genetic and political techniques of biodiversity and conservation research distinguishing S. *hongkongensis*. I've done this in order to exemplify what I see as a particular conjunction of ecological and sociological discourses on uniqueness and endemism in 1990s Hong Kong. The genetic testing of S. *hongkongensis*, and its simultaneous demarcating from and linking to S. *sinensis*, was of a piece with other attempts to endemicize natural life in Hong Kong in the years before and after the political transition. Scientific techniques for producing environmental value themselves gained value from, and gave value to, efforts to distinguish Hong Kong from China in other material and epistemic realms: social scientific techniques for defining a Hong Kong a way of life, biopolitical rationales for limiting immigration from China, calls to preserve Hong Kong's uniqueness as a free-market Chinese city.

Juxtaposing these realms is not meant to define ecology. Construing their echoes through the particular ecological figure of specific life—and asking what it means for political notions of autonomy and freedom when the figure of life around which they are built has specificity rather than organismic vitality as its defining feature—just might enable new interpretations with anthropological and political value. At a minimum, it casts the politics of autonomy in late-1990s Hong Kong in a different light: it denaturalizes the notion of an endangered Hong Kong way of life; it offers a frame for drawing contrasts between Hong Kong's democracy movement and other postcolonial freedom struggles; and it reminds us that Hong Kong autonomy as an end in itself accomplishes only the continued existence of things as they are.

Questions of economic and political autonomy in Hong Kong have undergone significant change since I began my research. In the 1980s and 1990s, what most concerned business leaders in Hong Kong was the threat of economic and regulatory redefinition. Hong Kong's continued economic life was believed to hinge on the region's political and economic distinction from China. Lately, however, the economic lines have blurred, as Hong Kong business interests have sought increasingly to tap mainland capital. The Hong Kong government has released a new kind of tourist visa for mainland Chinese, who have come in unprecedented numbers—seven million in 2005—to tour Hong Kong's newly opened Disneyland, to shop in designer boutiques, and to stroll along the Avenue of the Stars, a section of scenic walkway along Victoria Harbour dedicated to Hong Kong film celebrities. A new Hong Kong, one whose life lies in linkages with China, particularly within a new economic unit, the Pearl River Delta Economic Zone, is in the making. The new tourist visas and the currency influx they bring are part of this; the rest is realized through new bridges, highways, tax breaks, and business agreements linking Hong Kong with other cities around the delta, along with factories—new and old—in places like Shenzhen and Dong-guan.[27] At the same time, the Hong Kong government continues to jockey for international recognition as an ideal gateway to China. In 2004, Hong Kong and mainland China signed the Closer Economic Partnership Agree-ment, a free trade deal that "offers favourable trading and investment con-ditions to Hong Kong–based companies interested in exporting goods, or expanding business, into Mainland China."[28] The ecological formulation of Hong Kong's life thus appears to be changing. Hong Kong's survival rests not on defending a specific niche against encroachment but in distribut-ing its economic lifeblood. Its life hinges not upon the clear delineation of a species' characteristic edges but in its manner of negotiating porosity. Meanwhile, in the sphere of environmental politics, problems of air quality have drawn renewed attention. These problems require grappling with both the permeability of borders and the striking differences of atmosphere lived by differently situated people in a not so singular Hong Kong.

These changes prompt some questions about the politics of democracy in Hong Kong. If in the late 1990s this politics went hand in hand with efforts to forestall economic absorption by the People's Republic of China, the democracy movement of today is slightly more separated from economic interests as Hong Kong's business sector increasingly, pragmatically, sees Hong Kong's future as being predicated on integration with China rather than separation. Will this enable questions of self-rule to distinguish them-selves from questions of Hong Kong's economic viability? If they do, will issues of social and economic inequity find space for articulation within the present democracy movement? Could flourishing for all, rather than defense of a vaunted Hong Kong species, come to be what democrats and environ-mentalists desire?

In Sai Ying Pun, an old neighborhood on the western side of Hong Kong Island famous for salted fish, a group of men played Chinese chess outside my apartment building during the evenings of the warmer months. Seated on old boxes, the players swore loudly as they slammed down their pieces or muttered to themselves, resting their fingers on the piece they were moving so they could survey the board. The older men hung around near the park in the afternoon, waiting for the games to start. Some of the younger men worked during the day in the shark fin stores on my street. I knew the games were about to begin whenever one of the men in the park dragged the wooden crates out from behind the public toilet to set them up as tables and chairs. Every night, the same men and similar curses. Perhaps, I allowed myself the fantasy, the same $20 bill passed from the losers to the winners as the games went on.

"Hoi mei laan," my grandmother remarked, when I told her what neighborhood I was living in. "Seafood Lane. That's what they used to call Sai Ying Pun when I lived in Hong Kong." The dried seafood products used to come by boat in bundles from Tai O but now more often arrive in cardboard boxes from all over Southeast Asia. Dried fish, scallops, shrimp, and mushrooms add salt and tang to the thick summer air, and they sweeten the dusty gusts of winter. Sai Ying Pun strikes your ears, eyes, and nose the moment you enter the neighborhood. Vendors and shop-keepers hail as you walk by, "*Wei*, going shopping?" "Hey, done with work already?"

The security guard in my building told me that our alley was no longer open to car traffic because the local shopkeepers had taken over the street. "*Yuht baa yuht do*, they *baa* more and more." *Baa* is a colloquial word, connot-ing something in the family of "occupy," "seize," or "hoard." A realtor used it to laud the benefits of living on the ground floor of a village house. An upstairs flat was good, he explained, because you'd have a balcony. But that was it. If, on the other hand you had a bottom-floor flat, you could baa as much of the front of the building as possible with your stuff. That is, unless someone tried to stop you. But they probably wouldn't. You could even put in a barbecue, he added with a grin. The shopkeepers had baa-ed the street with dried goods, gradually extending their dominion with spirals of salted fish and rows of salted duck legs arranged on greasy sheets of cardboard. The goods were sunned to make sure they were uniformly dry—a residual bit of moisture could lead a piece of fish to rot.

Jeunggwan! Down slams the round plastic piece. Checkmate.

One Sunday some policemen with blue sweaters and notebooks came. They cited the shopkeepers who were drying their fish and duck in the street. They took the cardboard flats, dismantled the tables and crates, and had a crew of six or seven women dressed in Environment, Food, and Hygiene uniforms throw them in a dump truck. In the park across from my building, people hovered over their blankets and belongings, keeping themselves and their possessions just far enough from the market where they slept to be safe from prosecution for squatting. Once the tables, flats, and crates had been discarded, a bright orange truck equipped with a water tank pulled up, and the workers from Environment, Food, and Hygiene hosed down the narrow street and gutters as the shopkeepers looked on, perturbed.

"They do this every week," said one of the older gentlemen, who was watching from his favorite bench in the park. He kept his eyes on the drama in the street, barely glancing toward me as he spoke.

"Do they always cite people?"

"No . . . but they're getting much tighter."

"Why?"

"How would I know?"

I tried my luck with someone else. "They're cleaning up the street," explained the fruit vendor on the corner as I handed her a coin for some oranges. "You didn't know? They're cleaning up." She poked her finger down at the crate of oranges and three-legged stool perched by her feet. "I have to be careful about not letting any of my stuff slip off the sidewalk and into the street, or they'll fine me too."

The police and street washers eventually left. The orange truck trundled off.

Minutes later, as if by magic, a flat of fresh cardboard appeared in the street. Then another. And another. By early afternoon, the alley teemed with flats of drying fish, the shopkeepers had returned to their shops, a new chess game had started, and the squatters had dragged their belongings back into their homes. I went home to write about it; I knew that friends would like the story. But soon, everyone knew, the truck would come back to clean up.

ARTICULATED KNOWLEDGES

"If the government builds an incinerator here, you'll see an increase in the levels of dioxin, without a doubt!"

Nobody notices the chilly January breeze sweeping off the coast of Hong Kong's Northwest New Territories through the village of Lung Kwu Tan. Forty men have gathered in the village hall at a hastily arranged horseshoe of tables. Down at the open end of the shoe stands the object of their rapt attention: a large white man with soft features gesticulating wildly under the glare of an overhead projector, thundering to punctuate the details in his slides. Next to him stands a Chinese man in a purple fleece vest who translates the English presentation into Cantonese.

"You see, the government wants you to be Hong Kong's toilet! But you can say *no!*"

At the back of the room stands a Chinese woman. Occasionally she crosses her arms over her red coat and mutters something to the fellow next to her, who obliges by fiddling with the video camera he has trained on the presentation. Seeing some of the men in attendance head outside to smoke, the woman follows and peppers them with questions: "What do you think of this plan? Do you object? What are your demands?"

• • •

Hong Kong's Environmental Protection Department (EPD) made a startling calculation sometime in the mid-1990s: if Hong Kong continued to produce solid waste at a rate comparable to that at which it had over the last several years, the territory's landfills would be gorged with garbage by 2015, five years sooner than anticipated. To extend the life of these all too finite landfills, the EPD offered a plan to minimize the volume of Hong Kong's municipal solid waste. Their proposed means for doing so? The construction of a new incinerator capable of burning three thousand to six thousand tons of municipal waste a day.

An incinerator might seem a strange choice for a government department charged with the environment's protection, but for Hong Kong planners the incinerator made perfect sense. One only had to look at it within the terms of Hong Kong's relatively new integrated waste management plan. While everyone might agree that it would be ideal to see the reduction of consumption, the reuse of commodities and resources, and the recycling of materials like paper, metals, glass, and certain plastics, an incinerator offered a way to make Hong Kong's garbage smaller even if Hong Kong people did not do their part in reducing, reusing, and recycling. Garbage could be burnt to ash, and as a bonus, newly available technologies could harness the energy released from its combustion. Not only would their new incinerator reduce the volume of Hong Kong's garbage, EPD officials argued, it would produce electricity. Four sites for the new incinerator were nominated, and the government contracted ECC, an international environmental consulting company, to assess them.

Many people intimately concerned with the proposal, however, suspected that only two of the four sites were under serious consideration—both in Hong Kong's Northwest New Territories, an area already "blessed" with a power station and a landfill. The new incinerator would most likely be sited nearby, laypeople and experts deduced, to lessen the infrastructural work required to reroute garbage trucks and to route garbage-derived electricity into the power grid.

Unfortunately, a village called Lung Kwu Tan was situated squarely between the two probable sites. One site lay just south of the village; the other just north, in the heart of another village, Ha Pak Nai. Lung Kwu Tan residents, who traced some five hundred years of ancestral history in their village and who were already unhappy about the smokestacks and solid waste nearby, were understandably displeased by the prospect of having yet another piece of Hong Kong's waste management infrastructure in their vicinity. And unlike their neighbors in Ha Pak Nai, who felt largely discouraged and fatalistic, some of the men from Lung Kwu Tan resolved to fight the proposal.

When I first started writing about this struggle, I depicted it as a politics of place: a political struggle over a specific geography, where that geography is understood to be a meaning-laden environment materialized through the cultural practices and memories of its residents.[1] The residents of Lung Kwu Tan certainly had a deep connection to the place where they lived and were trying to protect it: A man laughs as he tells me how he used to be afraid of the pink dolphins that swam just off the coast because the village elders had taught their children that the pale creatures were monsters fond of eating naughty boys and girls. An elderly woman points behind the village to explain that in the old days Lung Kwu Tan used to be up there, in the hills, but they moved down gradually as they engaged more in farming. The vegetables, she adds, were covered for a while with ash from the nearby cement factory's smokestacks. "It's better now, but imagine how dirty an incinerator would be!" People with a history of dwelling in a particular place were confronted with drastic change. They were doing what they could to fight for their way of life, for the shape of the landscape they lived in.

My impulse to write about Lung Kwu Tan in this way stemmed in part from the ethnographic desire that I mentioned in chapter 1, a desire concerned as much with the politics of Hong Kong's representation as with the significance of place in environmental politics. Doing so would have the advantage of highlighting the textures of life in Hong Kong and in Lung Kwu Tan that merit attention—textures that ask for tropes other than breathless flights through a postmodern metropolis, or timeworn juxtapositions of skyscrapers and rickshaws, briefcases and temples, high fashion and rubber flip-flops. Hong Kong and the places it comprises are mundane—unique and mundane, and precisely in their landscape's saturation with everyday practice and memory, they are precious.

But to write about the politics in the making in Lung Kwu Tan, I realized, I would also have to reckon with translocal processes. The future of a place was certainly at stake, but the politics in process were in significant ways not happening in Lung Kwu Tan at all.

My first encounter with Hong Kong's incinerator came through ECC. In late 1999 I had secured a working relationship with the consulting company. I worked twice a week at ECC preparing various documents and reports in order to gain a sense of the quotidian activity of environmental technocrats for hire. My main task was to help with the incinerator site assessments. By chance, I was also getting to know the staff at Hong Kong's relatively new

Greenpeace office, and one such staff member told me that Rupert Yu, a Greenpeace campaigner, was beginning to organize against all incineration in Hong Kong. With visions of gaining a complex view of the incinerator controversy, I asked Rupert for permission to study him and his work. Soon I followed him wherever his work took him, from NGO office to Legislative Council to junkyard to village, gradually finding myself inserted into a network of scientific and political practices—a network at some points already made and at other points in the process of becoming. The meeting described at the opening of this chapter was one crucial moment in this network's becoming. It is this emergent collaboration between Lung Kwu Tan and Greenpeace, as well as the negotiations of expertise it involved, that I wish to account for and reflect upon here.

Expertise, Particulars, and Questions of Form

Scholars in numerous disciplines seem to agree that science, politics, and the relations between the two must be rethought. Sociologists and historians of science have drawn attention to the social and political contexts of the production of scientific facts.[2] Anthropologists have highlighted not only the cultural conditions of science but also how technoscientific objects accrue different meanings across various contexts and refigure cultural, political, and social relations.[3] Some have been keen to document "alternative" epistemologies and "local knowledges,"[4] while others warned of the elitism promised by the increasing value placed on expertise in state decision-making.[5]

Undergirding all these critical approaches is a skepticism about the presumed universality of scientific knowledge claims, as well as a commitment to empowering particular forms of knowledge that are occluded by faith in that universality. Questions of the particular and universal preoccupy students of the political as well. This is true especially in considerations of political formations in the context of North–South or East–West encounters, where the specter of Eurocentrism looms over attempts to establish unifying principles.[6]

Cultural anthropologists have typically assumed the role as the advocate or spokesperson of the particular in such discussions, amplifying the caution against the occlusion of cultural difference and specificity.[7] In fact, we often help to make particularity salient through the grounded density of our descriptions. This has been the case in our studies of both knowledge production and political movements. The tendency has been to increase analytic complexity by highlighting historical and cultural specificities.

In this chapter, I suggest that a different mode of anthropological engagement is necessary. Our concepts of "universality" and "particularity"—so crucial not only to anthropology but also to political theory and critical science studies—reach a point of analytic failure when confronted with transnational environmental politics. In such environmental arenas, these concepts are in a constant process of self-conscious deployment, production, and articulation.[8] This is nowhere more evident than in environmental controversies in postcolonial states, where arguments for and against the universalism and particularism of both political commitments, such as "environmentalism," and knowledge forms, such as "science" and "local knowledge," are always already in play.

To develop an analysis of emergent particularity and universality, I focus in this chapter on the negotiation and production of expertise in Hong Kong's incinerator controversy and the Greenpeace–Lung Kwu Tan collaboration. My main concern in studying this concrete enmeshment of the scientific and political is a question of form and form's effect. Questions of form and politics are of course familiar to science studies and to anthropology, but this chapter does not trace the rhetorics or controversies behind specific facts—about incineration, dioxin, or waste management, for instance—nor does it compare the efficacy of different technical solutions. Nor, for that matter, is its goal to contextualize environmental struggle, historically, culturally, or politically in Hong Kong. These are important analytic projects, but engaging them requires a different scaling of the problem than the one I choose here. In this chapter, I seek primarily to analyze the formal configuration of the environmental political field and to trace some of the analytic and political implications of that form's dominance. It is precisely by refraining from arguing the particularity of context that we might gain a crucial vantage from which to assess how particularity comes to work as a mark of expertise.

My argument proceeds in three parts. First, I note a paradoxical ecology of universality and particularity in the politics of environmental expertise in Hong Kong. While scientific knowledge is often associated with universality—or aspirations to universality—by both its critics and its advocates, a saturation of Hong Kong's political arenas with both universalizing and nativizing discourses yields a situation where, to be credible, expertise must bear universalizing and particularizing marks simultaneously. Thus, I suggest, the articulated knowledge bloc of the Greenpeace–Lung Kwu Tan collaboration constitutes not only a reply in content to the claims of Hong Kong's experts but also, equally significantly, a reply in and through form. Second, to illuminate how knowledges become articulated as counterexpertise, I offer a close analysis of translation in the collaboration, including an account of translation's pragmatic and metapragmatic effects. The account leads me to favor an understanding of both particularity and universality as concepts produced through, rather than preceding, political action. Finally, I explore the outsides of articulated knowledges, the presence of "unarticulated knowledges," emphasizing the political-economic conditions that have enabled only some people—and what they know—to count as the particular counterpoint to global environmentalism's universality.

If this chapter's argument concerns form, so too is its form part of its argument. The arguments and ethnographic descriptions in the following pages require and build upon each other; narratives of moments where expertise and knowledge are circulated occasion theoretical reflections on the nature and effects of such circulations, which in turn lead to other questions, other scenes, and their theoretical extensions. I thus seek to exemplify an entanglement of the concrete and the conceptual, and an oscillation between specifying and generalizing gestures, akin to those I describe and analyze.

Enter the Experts

"Ask them if they raise any chickens or pigs."

The English words echo off bare walls and desks, with an accent betraying a childhood spent in England. I am in Lung Kwu Tan, a day before the public town meeting. There are six of us today, all men, seated in desk chairs in the village office.

The command is directed at Greenpeace campaigner Rupert Yu who will act as the interpreter in tomorrow's presentation. Rupert heads Greenpeace China's campaign to restructure the way Hong Kong manages its trash. In particular, the organization opposes government plans to incinerate the bulk of its trash, citing fears that incinerators routinely emit dioxins—a family of possibly carcinogenic organic chemical compounds produced during the combustion of chlorine—into Hong Kong's already heavily polluted air. Rupert hunches intently over his knees, his arms gesturing vigorously as he asks in Cantonese, "Paul wants to know whether you raise any chickens or pigs . . ."

Rupert directs his translation to the three men from Lung Kwu Tan. Listening very intently is Jimmy, a tall handsome man in a blue collared sweatshirt and jeans. Jimmy is Rupert's main contact and collaborator in the village. Behind Jimmy sits Harold, a shorter man fond of wearing sports jackets. I recognize Harold; we met two years ago when I came to Lung Kwu Tan with members of Green Power, a local environmental organization. At that time, Green Power was working to forge a collaboration with Lung Kwu Tan to make the village over into a pink-dolphin ecotourist destination. Harold was one of Green Power's important contacts. In another corner sits a bigger fellow named Samson. He seems to have read all the recent newspaper articles about incineration and dioxins, and a few minutes ago he asked Rupert a question about Greenpeace's ship *Rainbow Warrior*. Like most male residents of indigenous villages in Hong Kong, Jimmy, Harold, and Samson share a surname—in this case, Lau.

Samson is a large man, but larger still is the white man in khakis and blue cowl-necked sweater who wants to know about pigs and chickens. Paul Connett, a friendly, gregarious professor of chemistry at Saint Lawrence University in New York, speaks at length about the possible effects of dioxin emissions on chickens and other livestock. Dioxins fall to the ground, he explains, and are then taken up by plants, which are then eaten by livestock, which are in turn consumed by people. Paul speaks easily and comfortably; he has been talking about these issues for years.

Originally from England, Paul received his bachelor's degree in natural science from Cambridge University, later moving to the United States to complete his doctorate in chemistry at Dartmouth College, and eventually taking up a teaching post at Saint Lawrence. He researches the effects of organic chemical compounds on biological systems, focusing on dioxins, and with his wife, Ellen, edits a bimonthly journal, *Waste Not*, which extols the benefits of reducing waste. Paul is a jocular fellow and fond of telling a good story. His large hands move with even more animation than Rupert's when he talks, at times cradling an imaginary piece of garbage, at other times forming benzene rings and dioxin molecules with his soft fingers. At the meeting he presses the men from Lung Kwu Tan for background information to better tailor his presentation to their needs.

Paul has made a passionate hobby of opposing incineration since the mid-1980s, when he successfully fought the construction of an incinerator in upstate New York. Then, as now, he was called upon for help because of his knowledge of chemistry. Paul has traveled widely to share his views and to support people who are fighting against plans to build incinerators in their communities. Before coming to Lung Kwu Tan, he had already given roughly sixteen hundred talks on these matters at speaking engagements in forty different countries and forty-nine of the fifty U.S. states, usually for Greenpeace International. He has come to Hong Kong at Rupert's request to lend scientific weight to the Greenpeace campaign.

Expert Witnesses

Witnesses figure prominently in the history of science. Some would even say that witnesses and practices of witnessing were crucial to the birth of experimental science as we know it. A classic text by Steven Shapin and Simon Schaffer documents the beginnings of experimental science in seventeenth-century England, focusing on a competition between Robert Boyle and Thomas Hobbes, both of whom sought to build an air pump capable of completely emptying a vessel of air.[9] The ability to make a real vacuum was a prize, for it would allow facts about air to be determined through comparative experiments. Shapin and Schaffer highlight the fact that Boyle, who eventually won the contest, staged demonstrations of his pump for the public — a public, that is, of upper-class men. Boyle's reasoning was that reliable men could attest to the soundness of his methods and the correct functioning of his vacuum. These men would serve as witnesses who could validate Boyle's experiments and his claims about what his experiments demonstrated. The authors use this, and Boyle's eventual success, to bolster their argument that experimental science's origins were social through and through.

While grateful for this point, Donna Haraway suggests that it's not sufficient to see these experiments as simply being staged within a preexisting social milieu.[10] New, modest, gentlemanly modes of masculinity — the very qualification of the men who validated Boyle's experiments — were themselves being made through these experiments, as much as the vacuums and pumps were. The fledgling experiments, in other words, not only yielded experimental science, they produced and hardened a conception of appropriate upper-class masculinity premised on restraint, civility, and unbiased knowledge (not to mention an ability to suppress squeamishness when watching birds die in glass jars evacuated of air). If today we assume that knowledge is best when its spokesperson is no one in particular,[11] this is partly because we have inherited and internalized the values for modest, stoic, invisible authority cultivated and propagated by Boyle and his witnesses. Haraway suggests recuperating the figure of the modest witness by rethinking what counts as true modesty—from an artificial and privileged invisibility to an avowal of social location and stake.

The international consultant of today is the twenty-first-century incarnation of the falsely modest witness. As international environmental regulations become increasingly stringent, developers in Hong Kong and elsewhere require experts familiar with these regulations, and with appropriate technologies, to help them comply with the new norms. The worldwide demand for expertise has proven to be a rich feeding ground for the relatively new field of environmental consulting. In Hong Kong the work done under the name of environmental consulting is varied. Some work looks like civil engineering—environmental consultants design plants, wastewater treatment facilities, plumbing, and sewage plans for large and small developments. Some of the work is research—consultants produce the environmental impact assessments (EIAs), now required for all developments in Hong Kong, that attest to the environmental acceptability of any development plan. And some of the work, present in most contracts, consists of attending public consultations and government meetings to answer questions or address objections on behalf of clients. Consultants serve, in other words, as expert witnesses and do more than observe and verify: they also testify.[12]

Yet the act of expert witnessing, entangled in the business of consulting, is clearly complicated by capital. Clients pay environmental consultants for their witnessing appearances. For this reason, it becomes difficult to discern whether the role of the hired environmental consultant is to offer expert testimony or to supply expert advocacy. An environmental consultant based in Hong Kong drew a provocative analogy to describe his role: "We're just like lawyers, only with science. A client hires you, and you argue their case. But we use science rather than the law."[13] Scholars in science studies could not sketch a relationship between science, rhetoric, and persuasion more clearly than this engineer did. Activists are, of course, quick to impugn the impartiality of industry-hired expert witnesses, even while they have consultants of their own.

Although the role of paid advocate is widely recognized and acknowledged, not all consultants assume it wholeheartedly. "Our job is to present the client with all the information so they can make good choices," maintained a senior consultant from an American consulting firm. "We wouldn't be doing our job if we just said things were okay. That would be a disservice."[14] In the face of cynicism persists the ideal of facilitating scientifically informed decisions.

If Boyle's witnesses needed to be modest and civil, the witnesses in the incinerator debate need to be expert. What counts as expert, though, is very much in the making in this process — made through the industry of international consulting, through the exclusion of nontechnical discourse in legislative meetings, and through the acts of testifying and speaking. Just as Boyle's air pump demonstrations served as technologies producing relations of gender and class, so do the meetings and testimonies in Hong Kong serve to produce relations between expert and lay knowledge, international and local location. What is made to count as expertise in these arenas is a paradoxical blend of particularism and universalism.

Expertise's Paradoxical Form

At the meeting in Lung Kwu Tan, Paul explains how dioxins accumulate first in the food chain, then in the body fat of human beings. Dioxins, he states, are so similar in structure to hormones present in a pregnant woman's body that they might disrupt natural processes during fetal development. Because of the work of environmental consultants such as those invoked by the EPD officials, Rupert needs Paul. Paul serves Rupert, Greenpeace, and the residents of Lung Kwu Tan as an expert witness and consultant. Unlike the EPD's consultants, Paul is not paid for his time. Nonetheless, he provides Rupert and the others with the same type of outside expertise, credible opinion, and experienced advocacy that industry consultants provide their clients.

Paul also works to meet a set of paradoxical requirements to qualify as an appropriate expert. This paradoxical configuration became clear to me at a special environmental panel meeting of Hong Kong's Legislative Council (LegCo). Convened a month prior to Paul's visit, the panel was charged with discussing the possibility of reactivating a medical waste incinerator in Hong Kong. Representatives from the EPD and Greenpeace were invited to circulate documents beforehand explaining their arguments for and against the proposal. At the meeting, officials from the EPD charged that the argument in Rupert's report, which outlined the health risks of medical waste incineration, was spurious. Despite having combed through countless tedious government documents about Hong Kong's waste management solution, Rupert was not considered by the government to be knowledgeable enough.

Although he had familiarized himself with the issues of transportation and refuse transfer; had learned to distinguish between municipal solid waste, clinical waste, commercial waste, and construction and demolition waste; and had educated himself about the chemical reactions that occur when garbage is combusted, the EPD officers and LegCo members did not consider Rupert an expert in incineration matters.

What struck me most was the EPD officials' reasoning behind their dismissal of Greenpeace's report. John Rockey, the EPD's director of waste facilities, and Steve Barclay, the principal assistant secretary of the Environment and Food Bureau, pleaded their case to the LegCo members in an idiom of locality. Greenpeace's conclusions, they argued, were based on data culled from the United States and did not adequately address the situation in Hong Kong. For instance, Hong Kong rubbish might have a completely different balance of plastics, paper, and wet organic matter than trash in the United States and would therefore yield different emissions when burned. Greenpeace was an outsider proposing solutions and identifying problems inappropriate to Hong Kong. Hong Kong needed a waste solution suitable for Hong Kong—and Greenpeace, a "U.S.-biased" organization, simply could not provide that.

The desire to develop locally appropriate technologies seems like a good thing. Steven Kurzman, an anthropologically trained user-interface designer, has taught me of a paradigmatic case illustrating the need for technological solutions that are locally appropriate. His guiding example is the case of prosthetic technology. Development NGOs, Kurzman argues, have a great deal to learn from Indian and Cambodian prosthetists, who design artificial legs made from readily available materials that are well suited to both regional climate and cultural norms of sitting and walking. American prostheses like the ones often disseminated by NGOs to South and Southeast Asia, on the other hand, dissolve in the humidity, break easily when worn without shoes, fall irreversibly into disrepair, and cause recipients to tumble when squatting. They exemplify the problems that come with assuming the universal applicability of Western technical solutions.[15]

Still, Rockey's and Barclay's criticism that Greenpeace derived their conclusions from U.S. data, and their plea for a waste management plan suitable for Hong Kong, made for a strange moment. Their arguments echoed the critiques of U.S.-centric thinking voiced by Kurzman and public intellectuals such as Anil Agarwal and Sunita Narain, who trenchantly analyzed the uncritical global application of Western environmental statistics and norms in the early 1990s.[16] Rockey and Barclay polemically posed a Hong Kong way against a Western way. That two British expatriates—who had served in Hong Kong's colonial government and kept their posts after the transfer of sovereignty—were speaking out against Greenpeace's supposed cultural imperialism, seemed strange indeed.

Complicating matters further was the way Rockey and Barclay proposed to defend their "locally appropriate" incinerator. An "eminent professor," Rockey explained, would appear before the LegCo to testify to the appropriateness and safety of the EPD's proposal to incinerate. Contrary to what one might expect from the localizing train of the EPD's arguments, however, this expert would not be a Hong Kong–based researcher. Instead of posing an unassailably "local" opposition to what they depicted as an outside, U.S. vision, Hong Kong's EPD officials announced that they had on their side an "international consultant." A consultant would secure their authority to solve Hong Kong's problems, and to decolonize Hong Kong environmentalism, in one fell swoop.

I highlight this turn of the EPD's argument not to criticize its contradictions but to explore what it indicates to be conceptually fundamental to the normative politics of expertise. Truth must scale down—particularize—at the same time as it scales up—universalize. In the power-knowledge field cleared through EPD's dismissal of Rupert's report and triumphant announcement of its international consultant, neither universality nor particularity is hegemonic. Universality is suspect (the apparently universal facts of Greenpeace's report are actually specific, "U.S.-centric") at the same time that particularity is insufficiently authoritative (the EPD's expert will be an "international" scientist). The EPD seeks a simultaneously particular and universal high ground. Expertise, according to the terms established in the meeting, must meet the postcolonial requirement of local appropriateness: science does not leave the building, it simply has to demonstrate that it has particular reasons to stay. A critical anthropology of expertise thus cannot rest on arguments about the situatedness of scientific claims—not because such arguments are too relativist to be productive but rather because the space of that critique is already occupied by the state.

So what has been made to count as expert in the context of the LegCo meeting? Circulation continues to be a crucial step in the valorization of expert knowledge. Expertise that is international and mobile—the esteemed "international consultant"—is posed against the knowledge of those who stay put. Like the ornaments that cycle through the Trobriand kula ring, knowledge gains value (as expertise) through its circulation and must continue to move across domains if it is to retain its patina. But movement is not enough; expertise must not only circulate but land as well. Expertise claims both universality and particularity—now counterknowledge must as well.

Accounting for Local Appropriateness

Here we have an interesting moment of accountability. The EPD, called to account by Greenpeace for the concrete environmental damage that may attend its proposed incinerator, in turn calls Greenpeace to account for a lack of specificity. This happens at a meeting that itself is meant to increase government accountability through transparency—the public consultation.[17]

This moment has several discrete conditions of possibility, which I suggest we think of as different strands of accountability permeating environmental fields. These strands, while originating from different contexts, converge here in a particularly potent way. The first, as Jasanoff and others have shown, is an emergent norm in global environmental governance that explicitly carves out a space for "local knowledge." This norm was institutionalized powerfully in the 1992 United Nations Conference on Environment and Development—often known simply as the Earth Summit—when drafters of various documents, such as the Convention on Biological Diversity, the Forest Principles, and the Rio Declaration, were careful to make explicit reference to the importance of "indigenous" and "traditional ecological knowledge," even creating an acronym (TEK) to anchor the bureaucratic reality of "traditional ecological knowledge."[18]

The architects of global environmental civil society did not come up with the idea of such a regulatory norm by themselves. Its necessity was made clear through ongoing grassroots and knowledge-making efforts of environmentalists and various communities, and the kinds of accountability they demanded. One of these, the second strand of accountability I wish to highlight, is the so-called Southern critique of global environmentalism. The Southern critique called "global environmentalism" to account for its unreflective application and universalism.[19] The third strand comes from the efforts of laypeople and activists to broaden what counts as environmental expertise through notions of citizen expertise and participatory research.[20] What needs to be accounted for in the second and third strand are geopolitical difference and the experience of "the people," so to speak—where "the people" might variously be people who burn fossil fuels to subsist, or people exposed to toxic emissions.

These strands do not map neatly onto the case at hand. Instead, they constitute a repertoire of forms shaping the claims made, demands voiced, and positions staked, as well as the contexts of political speech. The public consultation is not simply a context; it is a technique of government transparency and a formal solicitation of participation. Through it, the EPD is brought to account for the environmental harm that will potentially attend its incinerator. Through the consultation, Greenpeace raises the question of dioxin, through comparison with incinerators and their emissions in other places, and the moment of potential environmental accountability becomes actualized as a concrete concern about cancer risks and reproductive disorders. In response, the EPD calls Greenpeace and its reports to task for coming from, and invoking cases from, beyond Hong Kong's borders; in doing so, the EPD makes the specific environmental risks of dioxins identified by Greenpeace appear not specific enough. The EPD's demand for local appropriateness draws weight from the Southern environmentalist demand that global environmental advocates account for their global vision's blind spots; it also traces the contours of existing critiques of NGO–grassroots alliances.[21]

Perhaps we can be more specific than saying that the questioning of the universal is incorporated or voiced by the state. The questioning of universal claims (and the incitement therewith to account for particularity—of context, of economy, of trash composition) has been built into the systems and techniques of environmental, governmental, and nongovernmental accountability. These systems and techniques of accountability, in tandem with the positing of scientific expertise as universal knowledge, condition the require-ability (and only later, the requirement) of a universal/particular form for environmental expertise.

Form's Answer: Translating and Articulating Knowledge

If the universality of circulation and the particularity of local appropriateness are the conditions for expertise, how are these conditions met in the Lung Kwu Tan collaboration? To address this question, I offer a close account of translation. While the term's association with the domain of language will prove helpful—as expertise in my story circulates largely through various speech acts and speaking events—the concept also has material ramifications. Furthermore, translation not only produces knowledge as expertise (through circulation and repetition) but also animates contingent collaborations or "articulations" of knowledge that answer the formal requirement of simultaneous universality and particularity.[22] Assessing the politics of knowledge with attention to articulated and unarticulated knowledges can illuminate how some counterknowledges succeed when others fail.

Scanning the crowd at the town meeting, one sees an American chemist, a Hong Kong activist, an audience of village residents, a journalist, and an anthropologist. Rupert pushes for sustainability; incineration will only postpone the depletion of landfills in the medium term. Paul does likewise, also citing the health dangers of the dioxins, furans, and mercury that accumulate during waste combustion and ash disposal. Jimmy, a village leader, stands up to criticize the Hong Kong government for turning a blind eye to the illegal burning of trash in a nearby cement factory. Another, Harold, describes how ashes from the power station's smokestacks used to fall on vegetable plots. Still another asks why all of Hong Kong's garbage seems to be coming to Lung Kwu Tan.

Mixing the idioms of environmental organizing and feminist science studies, one might say that various stakeholders are present here, each with access to a differently situated knowledge of the problem at hand.[23] Rather than starting analytically with the people in the room, however, rather than taking for granted the sphere in which these differently situated knowledges meet, I want to ask: What brought these actors to occupy the same room at the same time? What practices allowed them to agree that they were talking about the same thing?[24]

Many researchers in science studies have examined the production of scientific facts through a framework of "translation," a framework attentive to the chains of translation and transformation through which scientists turn ambiguous materials into stable, unequivocal data, and through which actants enroll one another in their projects, transforming each other in the process.[25] Anthropologists have added complexity to this framework and to science studies at large by pressing for an account of translation and mistranslation in science's supposed periphery. Stacey Langwick and Stacy Pigg, for instance, mark the semiotic slippages between biomedical and lay idioms in Tanzania and Nepal, respectively, and emphasize the work it takes to produce the sensation of commensurability when vocabularies and bodily ontologies differ radically. They also raise the crucial question of what ideas, experiences, and even materialities might be lost in the drive to translate.[26]

My analysis here is indebted to these accounts and like them it concerns a moment of translation in the "periphery." Though one could fruitfully track the slippages in the translation in Lung Kwu Tan, I do not intend to make an argument about semiotic or bodily incommensurability. Instead, I am concerned with translation's metadiscursive role in political mobilization. To understand this role, I approach translation through a scale of analysis broader than the semiotic yet narrower than the historical; I am interested in translation's function as a speech event whose form generates metapragmatic effects in the Lung Kwu Tan meeting.[27]

Watch Rupert. Watch him glance at Paul while the chemist lists the various chemicals found in incinerator ash, then watch and listen as Rupert simultaneously plays thespian and interpreter, repeating Paul's speech in Cantonese, mimicking Paul's gestures, and infusing his speech with similar emphatic modulations of volume and pitch. Notice how he takes care to translate dioxin as *yiokying*, a Cantonese transliteration of the English chemical term that cleverly uses the word for "two" as its first character to capture the "di-" in dioxin.

Let us take this moment as a "mediational speech event." The term comes from the work of Richard Bauman, a linguistic anthropologist who argues that ritualized mediational speech events — events in which an intermediary speaker repeats the words of an original speaker to an audience — produce a number of powerful extralinguistic effects. Crucial among these is the generation of authority for the original speaker. Bauman observes, for instance, how the ritualized repetition of a chief's speech by his spokesperson imbues both the speech and the chief with greater power.[28]

Bauman's analysis of the extralinguistic effects of mediational speech is generative if we keep it in mind as we consider the meeting in Lung Kwu Tan.[29] For what is this meeting — with its clearly understood expectations that Rupert must repeat Paul's meaning faithfully — if not a mediational speech event? Authority for Paul is effected through the act of Rupert's translation, through that act's conventionalized respect for, and repetition of, Paul's words.

Furthermore, the repetition of Paul's English in Cantonese performs the movement of his know-how from the United States to Lung Kwu Tan. If expert value is attributed to knowledge that moves from one domain to another, then translation is one technology that makes knowledge move and come to matter as expertise. Certainly this is true in a straightforward way. Translation, in its conventional sense, allows knowledge to be communicated from one person to another even though they speak different languages. If, however, we assess translation not simply as a communicative act but as a mediational speech event, we glean another insight. As Bauman helps us see, the almost ritualistic repetition of the original speaker's speech in an act of translation itself generates authority.

Additionally, it matters that a scene of translation involves different languages; the scene is characterized not only by repetition but also by difference. The fact of difference is what allows the act of translation to serve as a kind of proof that circulation—expertise's precondition—is in process. While translation's explicit purpose is to move meanings from one semiotic world to another, that movement always presupposes and asserts through implicature an original distance between those worlds. The translation event thus not only moves ideas; it stages Paul and his words as being on the move and, by virtue of that movement, as expert. Translation performs transportation.

Translation, through the form of its event, thus generates the authority and circulations necessary for knowledge to count as expertise. It also enables articulation, or contingent unification across difference. To see how, let us turn once more to the meeting.

• • •

Paul addresses his audience, "In the United States, after coal burning, trash incineration is the major source of mercury." He pauses conscientiously so that Rupert can keep up with his Cantonese translation.

"Now the problem is not the mercury you breathe," Paul continues, "but the mercury that falls into ponds and lakes. When it gets into ponds and lakes, into sediment, it's converted into methyl mercury. . . ."

Paul stops again, and Rupert translates, "The problem is not the mercury you breathe. The problem is that when the mercury is spewed into the sky, it falls into the ocean, and the mercury will slowly sink into the seabed and become organic mercury." Rupert uses the Cantonese word *yaukee* to modify "mercury." In organic chemistry, yaukee typically translates "organic," and in Rupert's usage it technically denotes that elemental mercury has attached to another molecule.

Paul goes on, "Once it's in the form of methyl mercury, it accumulates in the aquatic creatures, with the concentrations increasing up the food chain, so that the big fish are a problem to eat."

Rupert translates, "When the organic mercury is in the seabed, then the small sea creatures will absorb the organic mercury. Then, when big fish eat small fish, and even bigger fish eat big fish, it will continue to be retained in the body."

· · ·

Rupert modifies and edits, depending on his own linguistic abilities and his assumptions about what his audience will understand, and he generates new meanings in the process. He explains Paul's "food chain" in steps: big fish eat small fish, and even bigger fish eat big fish. Rupert's work is fundamental to the building of the collaboration. More than a linguistic conduit, Rupert does the work of making Paul relevant to Lung Kwu Tan.

It is this mundane work of relevance making that calls to mind the concept of articulation. By articulation, I mean a linking together and enunciation of relevance between disparate elements. The utility of thinking political linkage and language together comes through clearly in a now famous metaphor from the cultural studies scholar, Stuart Hall:

> In England, the term has a nice double meaning because "articulate" means to utter, to speak forth, to be articulate. It carries that sense of language-ing, of expressing, etc. But we also speak of an "articulated" lorry (truck): a lorry where the front (cab) and back (trailer) can, but need not necessarily, be connected to one another. The two parts are connected to each other, but through a specific linkage, that can be broken. An articulation is thus the form of the connection that can make a unity of two different elements, under certain conditions. It is a linkage which is not necessary, determined, absolute and essential for all time.[30]

Articulation, in other words, is meant to capture a sense of the contingency of political collectivity, as well as the discursive conditions and work that make it possible. The concept can be traced from Antonio Gramsci's effort to theorize the formation of political blocs, through Ernesto Laclau's and Chantal Mouffe's work emphasizing the historical contingencies and discursive conditions for emergent political collectivity, and it has enjoyed a renaissance among anthropologists studying political mobilizations.[31] For students of environmental politics, the concept is useful for highlighting how environmentalist collaborations and politicizations of environmental materialities often emerge in tandem with new modes and terms for collective and subjective self-recognition.[32] Donald Moore, Jake Kosek, and Anand Pandian suggest that articulation offers a good general tool "for understanding emergent assemblages of institutions, apparatuses, practices, and discourses,"[33] and Tsing argues that the contingencies of articulation are good to think with when trying to understand ostensibly global formations, including global capitalism.[34]

As useful as the concept has been for framing linkages and assemblages, however, a question of process remains: how does articulation happen? Tania Li, an anthropologist of Indonesian indigenous politics, moves us close to answering this question. She compares two tribes in Indonesia that each sought to articulate an identity as "indigenous," noting that while one tribe succeeded, the other did not. Li concludes that a number of factors were at work—including the amount of urban interest in supporting the tribes' efforts, local political structures, and a capacity to present identity in forms intelligible to outsiders—but also warns that knowledge of such factors is of little help in predicting the outcome of an unpredictable process.[35]

Li's findings—particularly her conclusions that securing outside interest and intelligible communication with outsiders were crucial tasks—and my own experiences in Lung Kwu Tan suggest to me that translation practices are crucial to the process of articulation. True to the double significance of articulation, translation both voices claims and effects a kind of conjunction between domains that are not necessarily related. The act of translation reaches across distinct social worlds and asserts the relevance of one to the other.[36]

In this light, we see that Rupert's translation accomplishes several things. It facilitates communication between different groups; it generates authority for Paul through the mediational speech act form; it endows Paul with expertness, by performing through its structure the transportation and circulation crucial to present assumptions about expertise; and it establishes some of the necessary conditions for contingent collaboration between Lung Kwu Tan, Greenpeace, and Paul. Rupert's linguistic work makes Paul intelligible enough to the villagers that they will be interested in allying with him. The parties in the room oppose the incinerator for different reasons, but in occupying positions of difference relative to a common object, they become equivalent in Rupert's environmental translations.

This analysis of the tight relationship between translation and articulation resonates somewhat with an argument made by Judith Butler in the course of a debate with Laclau and Slavoj Žižek. She suggests that underlying the production of political universals is a process she calls "cultural translation."[37] In doing this, Butler not only emphasizes the work involved in expanding a political claim across contexts and the differences between the contexts that must be spanned, she also signals, through the inevitable question concerning the limits of translation, the gaps and representational violences that potentially attend any expansion of a claim's compass. Writing with her Left theorist interlocutors in mind, she thus enlists "cultural translation" primarily to stand for the importance of the concrete as opposed to theoretical conditions for making political claims. Cultural translation, Butler argues, is the concrete practice that moves aspiring universals into new cultural contexts, for both better and worse, leading to a general spread or transmission.

Looking at the moment of translation in Lung Kwu Tan shows, however, that translation accomplishes significantly more than transmission and spread.[38] On a semiotic level of analysis, translation is an act of meaning-transmission or meaning-negotiation. On a metapragmatic level, however, the translation event authorizes its source speech and, additionally, figures the source meaning as in-motion. In other words, translation not only spreads meaning but also, crucially, marks meaning with the sign—circulation—of the universal.

Equally germane to the question of universality is another metapragmatic point: that translation-as-event asserts through implicature a prior semiotic or cultural difference to be spanned. It draws the lines enabling the distinction and emergence of advocacy's different "enunciatory communities."[39] Thus, the translation event is paradoxically productive of difference, even while it builds a putative sameness across that difference. If the articulation of knowledges is indeed accomplished through translation, it might follow that articulation's power derives not from the unified voice it affords but from its constant implication of constituent difference. Translation qua articulation thus hails both particularity and universality and does so simultaneously. The political importance of articulation, then, is not that it spans particularities through the universal but that it enacts the co-presence of particular and universal interest.

With this co-presence, we are suddenly back to the paradoxical configuration of the technopolitical field outlined earlier. Articulated knowledges—

knowledges performatively scaled, linked, and mobilized through translation—are the political answer to the impossible challenge posed by expert regimes.

It should be clear at this point that to gloss the meeting in Lung Kwu Tan simply as a place where particular, situated knowledges met would be to miss a crucial piece of the picture. I have sought to demonstrate that these knowledges were not situated outside of, or previous to, the context of their articulation. This articulation was effected through acts and technologies of translation, and it generated the relationship in terms of which Lung Kwu Tan and Greenpeace and Rupert could find themselves (differently) situated at all. It generated that relationship as collaborative and collective. In generating the relationship, it also generated certain relational identities: mediating activist, expert, villager, male, local, American, and Chinese.

To some extent, the effort to articulate in Lung Kwu Tan was a success. Paul explained things basically but technically. He outlined the general schematics of an incinerator, explained the health risks associated with the dioxins and mercury that are often discharged from them, and compared incineration to waste reduction alternatives he had seen in his travels to other places. Jimmy from Lung Kwu Tan assumed the environmental mantle, intoning that this was not simply a Lung Kwu Tan problem but a Hong Kong one—we breathed the same air, after all, and ate the same food—and that Lung Kwu Tan would oppose the incinerator even if it were built elsewhere. Another man from Lung Kwu Tan offered to raid the cement factory after hours to procure a sample of the ash from the factory's on-site incinerator, and Rupert agreed to send it to Greenpeace's labs to test for chlorine. The villagers were ready to fight and even more eager than before to collaborate with Rupert and Greenpeace to draw attention to the scandal. Heady with adrenaline and with our own rebelliousness, we rearranged the tables afterward to enjoy a meal of puhntsoi, a dish typically reserved for special occasions composed of mixed meats served in a giant urn. Paul, a vegetarian, happily ate the chicken that someone planted in his bowl. Greenpeace's executive director, Chan Yiu Kwong, warned everyone not to overfeed Paul, as he already had "the runs"—whereupon everyone guffawed and someone thumped Paul affectionately on the back while Rupert explained to him what Chan had said. Meanwhile, Rupert and I followed Jimmy's and Samson's lead and dug our chopsticks into the various cuts of pork, beef, oysters, and shrimp piled on top of one another. A collaboration of activist men was born.

But not everyone was there.

Form's Limits: What Goes Unarticulated

"Say something! Say something!"

Flash back about fifteen minutes. Paul's presentation on dioxins and incinerators just ended, and several of the older men from Lung Kwu Tan are giving the *chunjeung* (village head) of Ha Pak Nai a hard time. Residents from both villages are mingling outside while Veronica Lee, a reporter, tries to gather testimonials for her documentary.

"Say something! What are you afraid of?" The men from Lung Kwu Tan can't understand why the chunjeung is so reluctant. He refuses to speak against the incinerator project or to testify to the impact that the nearby landfill has had on life in Ha Pak Nai.

Sullen, the chunjeung waves off his interrogators. "I'm not afraid. I'm just waiting for more *jiliu*, more data."

"Jiliu?!" laughs one of the men from Lung Kwu Tan. "You want jiliu? Tell them to come stand by the trucks. You can smell it! They can see the garbage juice leaking out of the trucks; they can see the flies. There's your jiliu!"

• • •

The chunjeung's refusal was not unexpected. The day before, Veronica had tried unsuccessfully to secure interviews from the chunjeung and others in Ha Pak Nai. Rupert, Paul, Chan, and I were seated in the village's general store. Nearby, a group of elderly men and women played a loud game of mahjong, and the storekeeper busied herself in the kitchen at the front of the shop, occasionally interrupting her activities to inject a word or two into the running commentary coming from the mahjong table. At the booth in the corner, near shelves of canned goods, sat the chunjeung and some other men from Ha Pak Nai, drinking tea and talking.

Veronica entered the store conspicuously, flanked by two men carrying a video camera and a boom microphone. The sounds of mahjong abruptly stopped, replaced by loud protests. Veronica seemed not to notice as she approached us. After exchanging pleasantries with Rupert and Paul, whom she had filmed earlier that day, she glanced at me. Had I come up with my story yet? No, not yet, I stammered. I was still observing.

At this point, she addressed some of the villagers. "Can I ask you a few questions?" People ignored her, and I noticed the chunjeung looked sour. Disappointed, Veronica turned to her cameraman and soundman, gave an almost imperceptible nod, and they suddenly aimed the camera and microphone at the chunjeung. "Sir, do you have anything you want to say on camera? What do you think of the government's plans to build an incinerator here?"

The chunjeung looked annoyed. "What's there to say? We don't have the materials, the jiliu . . ."

"I have the jiliu," Veronica countered.

The chunjeung scowled, then muttered, "Interview someone else." And with a pained look, he stood up and walked out the door.

Veronica and her crew turned to one of the companions he had left behind. "Did you know that jingfu [government] is thinking of building an incinerator here? Do you approve?"

"Ah, what's there to approve or not approve?" the man drawled. "If jingfu wants to build it, they'll build it, la."

The reporter tried the storekeeper next, but she simply shrugged her shoulders and said, "Ask the chunjeung."

Finally, Veronica bent down to address a very old man sitting on a stool next to mine. "Uncle? Uncle? Can you hear me?" But Uncle pointed to his ear, said he couldn't hear, and stared into his lap.

What do we make of refusals to speak? How do we understand refusals to articulate? A number of theories might explain why people in Ha Pak Nai chose not to talk to Veronica. One activist hypothesized that Ha Pak Nai was so poor that residents would be happy to be displaced; they could receive financial compensation or compensatory public housing. Another explanatory story might focus on Veronica's brusque manner as a reporter, which may have alienated her potential informants. "She wants an instant story, like instant noodles," opined an informant. Or perhaps they had given up and viewed the incinerator's construction as inevitable: "If jingfu wants to build it, they'll build it, la."

The chunjeung's own explanation could be found in an earlier conversation, during which Rupert and I had tried without success to convince him to have Ha Pak Nai join the Greenpeace–Lung Kwu Tan collaboration. Frustratingly noncommittal, the chunjeung told us that he had spoken up once in the past against a city plumbing project. But he and others from Ha Pak Nai had gotten their jiliu wrong and ended up looking like fools. So, he explained, he had to be very careful about getting his facts straight now, or he would risk losing more credibility.

At the time, and even now, his explanation struck me as evidence of the dominance of scientific expertise in environmental and planning controversies, and his reluctance to articulate spoke volumes about the marginal status of nonexpert knowledge in official processes of assessing the environmental and social impact of developments in Hong Kong. The chunjeung's recalcitrance after the meeting in Lung Kwu Tan could thus be seen as an indication of how familiar he was with the politics of expertise. Well aware of the strict criteria used to qualify knowledge as expertise worth listening to, he chose to remain silent rather than to speak without jiliu and be dismissed. The Ha Pak Nai deferrals and silences were, and were more than, refusals to articulate and collaborate. They illuminated, and evidenced recognition of, an unequally structured field for the politics of expertise.

Of course, it was obvious that beneath the chunjeung's silence lay knowledge. Consider how the men from Lung Kwu Tan teased him. They troubled his narrow definition of jiliu with their banter about how commonsensical the problem really was. Remember how they laughed, "You want jiliu? Tell them to come stand by the trucks. You can smell it! They can see the garbage juice leaking out of the trucks; they can see the flies. There's your jiliu!" Notice how their jibes manifested a maneuver of localization—they valorized their knowledge about the problems with garbage by valuing immediacy over jiliu. Their tropes of smell, sound, and sight located their discourse as more immediate, as closer to the problem, than government conceptions of jiliu. Through their teasing, they made the supposedly complex problem into a simple one, a matter that could be easily grasped by standing, smelling, and looking. They made it into a matter that the chunjeung knew plenty about.

Other people disturbed the notion of having jiliu as well. For instance, an elderly woman from Lung Kwu Tan first professed that she had little to say when I asked to interview her, claiming, "I don't know anything. I'm just an old woman, I don't know public speaking," but she soon launched into a rant: "The road, it stinks. They say that they're supposed to be washing it every week. Maybe, *maybe*, they come once a month. And the cars, the trucks that bring the garbage, they drive so fast. Last year, did you hear, there was an old woman from Lung Kwu Tan who was killed when a truck ran her over. It's not safe. The flies come. The smell. They put the power plant here. Then the landfill here. Why do they put everything with us?"[40] The woman continued to talk for over forty-five minutes. As she went on and on about the specifics of what it means to live next to the landfill—even while she professed to know nothing—I could not help but think that she was, in effect, talking over the divide distinguishing her knowledge from "expertise."

One might identify these events as instances of "back talk," moments in which people at the margins voice critiques of present structures of power.[41] It seems crucial to recognize agency and resistance, and I am sympathetic with this line of argument. If our aim, however, is to understand the mobilization of knowledge in the politics of expertise, we must also ask the difficult question of whether and when such back talk achieves political effect.

To focus on this question, it is helpful to recall the key components of both Laclau's and Stuart Hall's sense of "articulation." Articulations are both linguistic and connective; can we say the same about instances of back talk? They are certainly enunciative, but are they connective? Do they manage to enroll allies? Consider the incidents with Ha Pak Nai's chunjeung and the woman from Lung Kwu Tan. The chunjeung claimed not to have enough information, while everyone else thought he had plenty. The Lung Kwu Tan resident claimed an inability to speak publicly yet would not stop talking. What can a consideration of these anecdotes offer to a theory of articulated knowledges?

I understand these two incidents as moments of unarticulated knowledge. The subject of the first incident refused to speak, while the subject of the second professed simply to be uncomfortable with speaking. In both events, detailed accounts of experience seemed to contradict claims of knowing too little or having nothing to say. These anecdotes, if we do not take them as examples of delusions or deliberate diversions, can be understood as moments where subjects recognized the limited power of what they knew. Although jiliu was articulable and, indeed, was articulated in a linguistic sense by those who teased the chunjeung, and although the loquacious elderly woman was not nearly as inarticulate as she claimed to be, the knowledges voiced in these events remained isolated, unarticulated.

Let me be clear: a notion of unarticulated knowledges is not intended to deny the existence of knowledge or back talk, or eloquence in their production. Even if I call my conversation with the woman from Lung Kwu Tan a moment of inarticulateness, for instance, I recognize that she was an articulate and persuasive speaker. Yet she and I also recognized that however fluent it may have been, her speech to me was not "public speaking." She recognized that it did not engage, or forge connections with, a public. In this sense, in recognizing its inability to connect up with something larger, to mobilize and be mobilized in a collective, we can say her speech did not articulate.

A concept of unarticulated knowledges requires us to think about whether an instance of back talk—even if voiced in the most fluent of ways—"articulates" in the political sense, and to ask why it does or does not. It is less concerned with establishing the agency, knowledge, or eloquence of a speaker than it is with assessing the lay of the political and discursive terrain within which she expresses herself. Such an assessment requires adding to our formal analysis of the metapragmatics of articulation and translation a frank account of the political-economic specificities that shape the conditions of their felicity.[42]

• • •

"What can we do?" she asks. The woman I am talking with has lived in Ha Pak Nai for about forty years. She moved there from Yuen Long. "There's nothing we can do. If jingfu wants to build it, they'll just do it. Maybe they'll puih ngok, compensate for the houses. They should also puih tihn, reimburse for the fields. Look at these fields, all these trees I've planted. Don't you think that government should pay for this too?"

"But it's not like we're indigenous. Government would have to pay them for the land. By the foot! But we don't own the land. I pay rent for this land. But jingfu will have to worry when they encounter the land here that's owned by Ha Village. A lot of the land in Ha Pak Nai belongs to Ha Village. For example, I plant these trees, but I pay rent to Ha Village. Like a thousand dollars a month. Lung Kwu Tan, Ha Village, they're both indigenous. Government will have to pay them."[43]

• • •

The political optimism in Lung Kwu Tan and political pessimism in Ha Pak Nai hinge on the political economy of indigeneity. The term, "indigenous inhabitant," is a relative newcomer to Hong Kong law, adopted only in 1972, but it fills a space of governmental exception that can be traced back to 1898, when China ceded the area now known as Hong Kong's New Territories to Britain. In part to quiet resistance from villagers who feared the British would expropriate their land, in part because indirect rule was the method of choice for British colonialism, the colonial government promised residents of the New Territories that it would honor existing land rights and customary Chinese practices concerning the use, ownership, and transfer of land. The rural resistance quieted down, and the New Territories were ceded to the British without further ado. This concession of village lands required no great consideration on the part of the British at the time, as the villages were small and scattered, and were not considered strategically important.

This technique of rule had the requirement and effect of reifying custom and individuating ownership. Administration of colonial land policy began with a territory-wide survey to establish the ownership of particular plots in existing villages. Even when colonial authorities understood custom to be complex—such as the widely understood situation where different people might hold surface-soil and sub-soil rights for the "same" plot of land—only one owner was registered. The survey also drew a conceptual boundary distinguishing New Territories villages and residents who would have special rights under Hong Kong law to preserve their customs (and to argue for the "customary" nature of certain practices), from villages and people that did not have these rights.[44]

Today, indigenous status accorded to people whose ancestors inhabited villages that made the registry in 1898, in conjunction with capitalist meanings of land, has afforded significant privileges to residents of Lung Kwu Tan and other recognized villages. Thanks to the Small House Ordinance of 1972, male descendants in indigenous villages can quickly convert agricultural into developable land to build residences, circumventing other ordinances that slow down most Hong Kong developers.[45] That many villagers used the Small House Ordinance to build triplexes that could then be sold or let in pieces is not surprising given Hong Kong's lucrative real estate market. The public uproar over the privileges afforded by the ordinance and exploited by indigenous villagers continues to shape the public discourse on indigeneity in Hong Kong, all the while obscuring the fact that the rest of Hong Kong's land was quietly reserved for the state by the same ordinance.

Lung Kwu Tan residents, some of whom work in real estate, are thus enabled by the authority, rights, and economic privileges that come with their status as indigenous, and many have developed strong political connections. As it happens, one of the most prominent members of Hong Kong's Rural Committee—a powerful, largely pro-Beijing government body formed to represent rural and village interests in Hong Kong—hails from Lung Kwu Tan.

One of my informants reminds us that things are dramatically different in Ha Pak Nai. Indigenous villagers own land; people in her village do not.

Indigenous villagers earn rent; residents of her village pay rent to plant their fields and trees. Jingfu will have to pay the residents of Lung Kwu Tan dearly for land if it builds here; it most likely will not reimburse people in Ha Pak Nai whose fields are planted on that land. She shows us that the legacies of colonial recognition are lived today as differential rights and relationships with resources and as differently remembered histories of success and failure in negotiating with the state. I mobilize her reminder and the earlier moments of unarticulated knowledge to effect my last analytic articulation: the simultaneously general and specific concern that issues of political-economic inequality easily go unarticulated in the articulations of environmental expertise.

Implications

I end with implications. Let me begin by acknowledging the sensation one might have that there is nothing new to say here. The environmental justice literature has convincingly established that the siting of landfills, nuclear power plants, waste treatment facilities, and other sources of environmental harm enacts and exacerbates existing structural inequities, and that this is as true for garbage as it is for toxic waste.[46] Critical studies of expertise have identified the elitism of the environmental technocratic spheres and their elision of local knowledges at the expense of the supposedly global idioms of scientific expertise, and highlighted the struggles between representatives of the "local" and "global."[47] The stories relayed in this chapter, with their almost typical cast of characters—Greenpeace activists, traveling experts, government officials, local communities—might be seen as adding simply another piece of documentation to those well-established frameworks, even if the Hong Kong situation offers some local variation on the global theme. If so, the benefits of the present study are not analytic, only documentary. There are no surprises.

One possible response to this sensation is to deny it: to assert that the details of the collaboration in Lung Kwu Tan and in Hong Kong's Legislative Council exceed previous forms of argument and therefore require a fundamental reorientation of the theoretical apparatus. I could say, for instance, that the politics of expertise in Hong Kong is fundamentally different from those elsewhere, that this is evidenced when British expatriates argue in an idiom of locality, and that this is an outcome of a specific Hong Kong history. To do so would follow a method of defamiliarization familiar to anthropologists, though our call to make the familiar strange was historically a call to estrange our own cultural assumptions through comparison with otherness. A desire to assert the peculiarity of Hong Kong environmental politics, however, comes from a slightly different set of stakes. It is as if one cannot stand to see the supposed other looking analytically familiar. This analytic habit has the apparent benefit of guarding against the unreflective application of framework—by demanding a "locally appropriate" theoretical technology, as it were. Each locale has its own analysis, its own institutional and technological framework.

It should be clear by this point that I think we should be more reflective in our arguments for locally appropriate analysis, particularly when what seems particular or peculiar about the normative configuration of expertise in question is itself a certain call for local appropriateness. Our frameworks are implicated in each other. It is time to make strange this familiar mode of argument and to analyze its function and effects in the exercise of power in contemporary regimes of expertise.

For this reason, I avoided particularizing arguments and instead investigated the conceptual forms underlying the very idea of local appropriateness: a specific critique of universalism and a normative theory of simultaneous universality and particularity. I demonstrated how these forms are invoked in government, how they norm the politics of expertise such that the only viable shape for counterknowledge is one of articulated knowledges. I then analyzed the metapragmatic processes through which an activist's translations enabled an assumption of these forms for counter-expertise by an emergent collaboration. Articulated knowledges respond to the requirement that political knowledge claims to be both universal and particular simultaneously through these forms. My analysis of the anti-incinerator coalition thus did not seek to identify local knowledge or a locally appropriate solution but looked instead to the practices by which knowledges were differentially scaled, linked, and mobilized.

This insight about the mobilization of knowledge required a graft of political theory and science studies. I extended theories of political articulation with science studies' insights about the translations and collective formations that enable the making of knowledge. Simultaneously I reframed the production of authoritative knowledge within a context of political contest and coalition. Even so, articulation was not allowed to govern the analysis, for a metapragmatic analysis showed that translation, articulation's fundamental technique, enacts the simultaneous marks of universality and particularity. In other words, difference, universality, and particularity, rather than being preceding characters of entities that are subsequently linked politically, are in fact all characters metapragmatically generated and linked through translation's event. Through an assemblage of methods, then, I drew a different kind of interpretation of the familiar data—in fact, an analysis of the production of the data's very familiarity to us.

One analytic and political implication that seems clear is that the critique of universalism is as dead as universalism itself; the critique no longer speaks to the configuration of power but instead finds itself echoing the state. More mildly put, an indictment of universalism in expert venues is helpful, but not sufficient, for the critical analysis of expert politics. This critique—and its valorization of specificity—has already been internalized by the state and its experts.

While the new terrain for the politics of environmental expertise seems more commodious than the previous in its accommodation—indeed, its requirement—of particularity, what will stand for the particular in the making and declaration of particular-universal political claims is still to be determined. As we saw in Ha Pak Nai, histories of political-economic inequality and marginality might structure the exclusion of concrete concerns as much from particular-universal mobilizations as from more universal formulations of expertise. Particularity is thus nothing to rest on, analytically or politically. Rather, because particulars must always be mobilized and articulated to count as such, they must always beg the question of their insides and outsides, and the old and new power relationships shaping their articulate and inarticulate moments.

A long, red outline reaches toward an oval drawn in blue. In case you can't tell what they represent, the artist has carefully labeled each shape on the piece of A4 computer paper. Laughing, Maria removes her eight-year-old's drawing from her bulletin board to give me a closer look. "Look, can you believe it? I just told Bronwyn about sperm and eggs once, when she asked me where babies came from. I forgot about it, but I guess she didn't, because she just drew this at school yesterday! She's really, really smart, that girl!" She beams, then points suddenly to her bleached hair, "I wonder sometimes what the teachers and other parents over there think of me—I'm blonde, I work for Greenpeace, and my daughter's drawing pictures of sperm!" Her pride, palpable and glowing, is as much for Bronwyn's broaching of taboo topics as for her precociousness. Maria is an unorthodox woman, and signs point to Bronwyn being as exceptional as her mother.

When I think of Maria, I picture her in her favorite baggy overalls, eyes sparkling with mischief and wicked wit, her once black hair bleached three shades blonder than the light brown mop of her daughter. She's laughing and, always in my mind, she's talking. Her mastery of language astounds me. She's not just fluent—she's funny, smart, and eloquent in Cantonese, Mandarin, and English. She moves as easily through tiny restaurants in Sai Ying Pun, eating and talking over plastic-draped tables of clay pot rice, steamed scallops, roasted goose, or hot sesame pudding, as she does mingling over gin and tonics with expats and reporters at the Foreign Correspondents' Club in Central.

When Maria was first considered for a position as Greenpeace's press officer, Chan Yiu Kwong and Sarah—the woman from Greenpeace Vancouver who came to start the local office—asked around about her reputation at work. Maria was at that time a senior editor of Hong Kong's leading English-language newspaper, the South China Morning Post, an institution harking back to the region's early colonial years. She was one of the paper's only female senior staff.

Unfortunately, she had enemies. Colleagues who worked around and under her reported bad news to Yiu Kwong and Sarah: "She's a bitch! When she comes to work, she's a bitch. She doesn't budge!"

These reports concerned Yiu Kwong, Maria remembers. "Typical Chinese way of thinking," she says. "Typical to want a woman to just be nice and get along with everybody." He had been unsure after these reports, but Sarah had a completely different take: "Sarah thought, Great!"

"She's too tough, they'd argue—but the badmouthing was great!" Maria savors the thwarted betrayal. Rather than hearing signs of a flaw, Sarah heard the reporters' complaints as evidence of Maria's ability to stand her ground among unfriendly faces—a crucial capacity for the press officer of a fledgling activist organization. Yiu Kwong may have wanted Maria to be more gentle and easygoing, but the Canadian woman thought a tough bitch would be perfect for Greenpeace. So here she was.

Maria may scoff at what she calls typical Chinese culture, but she holds no illusions of cross-cultural feminist solidarity. The main reason she had been tough with colleagues at the Post, she explains, was that they wouldn't otherwise take a young Chinese woman seriously. White men, she remarked, seemed to have no problem with her. But a white woman, one whom she had actually helped to find a job at the paper, had bristled at Maria's leadership and led others against her.

Maria continues to be tough. She won't beg reporters to attend actions or press events. She tells them what Greenpeace intends; they can show up or not. On occasion, she says with a laugh, she's a bit bossy and tells reporters how to write their stories. And the reporters usually listen—they know she was a reporter and an editor, that she knows the best line for a story, that she has been the boss of some of them.

Yiu Kwong may have had doubts early on, but now he trusts and relies on Maria. And they're friends now. She likes teasing him, especially in the summer when he wears shorts. She tells him he has nice legs and that she really likes guys with hairy legs. This, she reveals, is the real reason she dates *gwailou*, or Westerners—because most of them have hairy legs—but his are nice and hairy too. Yiu Kwong laughs and backs away.

EARTHLY VOCATIONS

Rocks and Water

Riding Greenpeace's Zodiac, an inflatable raft propelled to high speeds by a 200-horsepower Honda outboard motor, felt even more thrilling than it looked on television. As we bounced over dark jade swells I looked all around me — happy to be on the water, disoriented to see Hong Kong's skyline from a distance, giddy to have sky above my head and to feel fast and free. From here, all the high-rises looked organic, like a concrete fungus dwarfed by the island's lush cliffs. I gulped greedy mouthfuls of clean air and held on tight.

We numbered eight that day. Around me sat my friends Wing Hung, Siuming, Faanshu, and Ginny, along with two of their good friends from university. Steering the raft, in yellow tinted sunglasses and trailing a flapping, loosely buttoned khaki shirt, was Rupert. No official Greenpeace business that day, but Rupert needed to take the Zodiac out for maintenance anyway, and he was happy to captain. "Is everyone okay?" he asked us from time to time, turning to check on his passengers. "Tim, okay? Siuming, okay? Good!" And with a firm nod, he would turn back to resume attention to driving.

Wing Hung and Faanshu, along with their friend Ah Guo, had organized our outing. Our mission was to investigate Penny's Bay, the site chosen for Hong Kong's new Disneyland. The government had recently arranged a HK$23 billion deal with the Walt Disney Company to build an amusement park on Lantau Island, drawing intense scrutiny from local activists and both local and international media. Tung Chee-hwa, Hong Kong's chief executive, hoped the park would help invigorate the economy, which had been struggling of late. Since the Asian currency crisis of 1997 and the bursting of the bubble generated by years of relying on the inflation of real estate values for economic growth, property values were down, the unemployment rate had risen to a figure nearly three times that of 1997 (6.2 percent in 1999 as opposed to 2.2 percent in 1997), and Hong Kongers—pundits and laypeople alike—had begun grumbling about the inadequacies of Hong Kong's first chief executive since its political handover to China. Tung responded with a barrage of ideas that ultimately fell flat. Hong Kong could become the world's herbal port, a gateway to import and synthesize Chinese medicines. A "Cyberport" proposal promised to establish Hong Kong as Asia's premier information technology hub. If those didn't seem outlandish enough, Tung now sought to bring Disney to Hong Kong, in hopes that much needed investment and tourist dollars would come with it and create thousands of jobs in the process.

Wing Hung invited me on this outing the night I arrived in Hong Kong, when he and Siuming greeted me at the apartment in Sai Ying Pun they had helped me find. What we were out to do exactly, I wasn't sure. Perhaps we had just come to confirm for ourselves the magnitude of the development that Tung Chee-hwa had approved. Ah Guo had brought a map, and when we found the site, he pointed out where the construction was most likely to begin. Developers planned to reclaim the whole bay, despite protests from the Antiquities and Monuments Office over the archaeological finds that had been dug up from the bay. The environmental costs looked extreme too. Like Hong Kong's airport, the park would require extensive dredging of harbor mud during construction; already scientists voiced concerns that drifting particles could affect local fish farms, and the mud on the ocean floor was suspected of containing toxic chemicals. Faanshu and Ah Guo had recently started an organization called Beware of Mickey! that criticized the project on cultural and economic grounds, and just a month ago a local labor rights group had publicly indicted the labor practices in factories making Disney souvenirs.

Before long, we climbed back into the Zodiac to look for a place to swim, ruling out one inlet where the waves had a foamy residue and anchoring in a small cove nearby. Most of our troupe were not experienced swimmers, so some floated on the surface wearing the life vests from the raft. Rupert, Siuming, Wing Hung, and I, though, splashed unencumbered.

"*Waah, hou man, ah!*" ("Wow, very 'man!'") a voice cried out. This phrase puzzled me. Earlier someone had uttered the same thing when I pulled my-self from the water back into the raft. This time the voice was Faanshu's, and his tone suggested three parts amazement and one part mockery. Look-ing over to where he was pointing, I saw Rupert bouldering—picking his barefoot way up the cliff near the shore. Rupert moved heavily but confi-dently, slapping each handhold several times with an open hand to test its strength before committing his weight to it and muscling his way up. We all laughed when he found himself wedged in a horizontal crack, looking like a trapped chubby worm. And we watched, rapt, until he extricated himself and climbed back down to safety.

The October heat was making me drowsy, so after diving a few times in the cooler, deeper water, I hoisted myself onto a small sun-warmed rock, ready to nap off the residue of transpacific jet lag. A few moments later, I heard a splash and opened my eyes to see Rupert clamber up beside me. We nodded to each other and looked at the cliffs for a while.

Rupert broke the silence in English. "So you're from America?"

I acknowledged I was, and he probed further. "Which coast?"

"The West Coast," I said. "California."

"Ah," Rupert looked satisfied. "The West Coast! My wife and I lived on the West Coast in Canada. In British Columbia.

"Oh, you're from Canada?"

No, he said. He had been born in Hong Kong, but he and his wife, May, had moved to Canada several years ago, and they had lived there for five years. "Then we moved back to Hong Kong to help start the Greenpeace office. It was great! I love the laid-back West Coast lifestyle. I much prefer life there. It suits me much better."

We watched Wing Hung and Siuming for a while. Wing Hung paddled gamely from a rock to Siuming, who was floating on her back under the bright blue sky after swimming a few hundred meters of freestyle. A bit farther off, Faanshu and Ginny paddled around in their life preservers.

I gestured to the rocks where Rupert had been bouldering earlier. "So, do you climb a lot?"

"No, not much anymore. My partner in Canada used to take me out all the time. We'd go out with foldable kayaks to spots that were really remote, and then climb. But in Hong Kong I haven't had much opportunity. I love it though. Do you climb?"

I told him I had begun climbing back at home, and that though I wasn't very experienced, I would be excited to do some outdoor climbing in Hong Kong. We agreed that it would be a great way to get outside and away from the city.

Meanwhile, I noticed that Wing Hung had come a long way with his swimming. I wondered how many other Greenpeace staff could swim. Before joining Greenpeace, I knew, Wing Hung couldn't swim a stroke. And someone had told me that Maria, the press officer, was afraid of water. She had gotten much better lately, though. Now she was even studying to captain the Zodiac.

Earthly Callings, Worldly Ethics

In chapter 4 we saw the central role of translation in the staging of environmentalist controversy, particularly how through its form and the metapragmatics of its event, it generates scales of particularity and universality, along with their geographic analogues, locality and globality—generating them only to performatively link them in political articulation. This chapter tackles an aspect of this process not yet addressed, namely the question of how environmentalists, the agents of such translation, come to be.

This question is often approached in environmental thought as a problem of awareness.[1] How do people come to be aware of and to care for the environment, nature, ecology, sustainability, resources, and the like? How can they be brought to care enough to act differently? But environmentalism, if we may name such a thing, is not simply about what one knows. It is, rather, a way of forming and situating oneself in the world—in relation to nature, to the planet, to science, to other human beings, and to other living and non-living forms.[2] Just as religions, medical regimes, and economies emerge as conjunctions of institutions, knowledges, and norms in which processes of subject-formation and practices of self-care enable their articulation, so too does environmentalism hinge upon the cultivated details of living. Earthly ethics are what make environmentalists and environmentalism real in the world. Sometimes, they even make swimmers.

This chapter offers an ethnographic account of some of the particular ways environmentalism came to be folded into and made manifest in ethical practices of self-making in Hong Kong, and how certain practices became so bound to some people's sense of themselves that they grounded action at odds with the order of things. In particular, we will see how two environmental professionals—Rupert Yu, the Greenpeace campaigner who organized the town meeting discussed in chapter 4, and William Lee, an environmental engineer at the Hong Kong office of a New Jersey–based environmental consulting company—came to recognize themselves as environmentalists, and what significances their particular ways of approaching environmentally inflected work had for them.

Rupert and William are not intended to be representative of Hong Kong environmentalism; they are too particular for that. But their stories and the ways they told them to me throw into relief a key and unavoidable issue at stake for people who engaged in environmentalist practices in Hong Kong in the late 1990s and early 2000s. In their lives, their work, and their declarations, Rupert and William negotiate the cosmopolitanism of their environmentalist commitments and their selves, linking and shuttling between Hong Kong and other places, just as Rupert and I did in our talk about rock climbing that sunny day. This tacking between places, I suggest, is what gives environmentalist action the traction that it has in these two subjects' lives, even while the ties that such tacking performs can become grounds for political suspicion, as we saw in chapter 4. A self-conscious ethic of environmentalism in Hong Kong—and what environmentalist ethic is not self-conscious?—is completely interarticulated, interwoven with other regimes of living, other modes of self-craft and care, and other moral ecologies. Environmentalist practices and meanings do not stand or work on their own; they overlap with other desires and meanings, including the meanings of location and of attachment to other places.

Cosmopolitanism, glossed by some commentators as a state of "thinking and feeling beyond the nation,"[3] is a politically fraught concept. In social theory, it cannot be separated from the notion of political universality, whether one turns to Kant's essay "Idea for a Universal History with a Cosmopolitan Intent," which propounded a philosophical recognition of universal human rights, or to Marx's later materialist argument that capital itself was cosmopolitan, a concrete universal in the historical present that required a cosmopolitan political consciousness insofar as workers needed to imagine solidarity with other workers in far-flung places.[4] As the grounds for political universalism have become increasingly suspect, so too has cosmopolitanism evolved from an idealized concept of supranational consciousness to a term that necessitates situated description. Business executives, traveling academics, wealthy immigrants, refugees, and migrant workers, for instance, arguably all think and feel beyond their geographic and geopolitical locations; but they and their politics trace different routes.[5]

Cosmopolitanism in a general sense, furthermore, can be entirely consonant with regional or ethnic exclusivity.[6] Thinking and feeling beyond the nation offers no guarantee of an all-inclusive conception of universal human community. And yet it might be precisely through a troubled sense of cosmopolitanism that we find political hope. Through situating cosmopolitanisms even as they are forged, political universals and the commitments they nurture might be more viable and more robust, precisely in accounting for their pathways and commitments.[7]

In this spirit, I offer a portrait of earthly vocations in Hong Kong. "Earthly vocations" is my shorthand for the simultaneously personal and political itineraries of living, traveling, and acting in relation to particular ideas of environmental protection, nature, and place. The ambiguities of the English word *vocation* are fortuitous; conjuring both theistic and secular senses, it evokes environmentalism as a kind of secular calling. At the same time, it points beyond issues of political belief or environmental awareness to the practical matters of living a life, not to mention the stakes of spreading the word.[8]

I broach this religious comparison uneasily, as I feel more adept at dealing with questions of science, politics, and culture. But if we wish to better understand how particular subjects come to be committed to particular scientific and political truths, how people come to be invested in translating those truths across social and cultural borders, and the violences and arrogances that attend the expansion of range for a particular truth, then an account of conviction and vocation is in order.[9]

Certainly, environmentalism begs for such an account. Its history tangles with religious histories and significances. Notions of environment, wilderness, and conservation emerged in a crucible of Christian asceticism, skepticism about modernity's promise, and frontier dreams of westward expansion, as well as violent erasures — symbolic and material — of American Indian life in what came to be known as America's "wilderness." Then there are the details, such as Alexander von Humboldt's breathless declarations of wilderness's sublimity, or John Muir's Calvinist upbringing and his praising of mountain valleys as natural cathedrals. These clues to religious sense and biography give flesh to the idea that the fervor of Humboldt, Muir, and those following them to impart the sacredness of nature to others has been missionary in more ways than one.[10]

Most striking to me in these histories are the central roles played by particular practices and pathways of movement in people's quests to transcend modern ills through nature. Environmentalist and Christian convictions are woven through itineraries of travel and retreat, cultivated sensibilities, and public declarations of nature's sanctity. These are earthly vocations indeed — not only for the earth, but lived in the world. To keep this in mind means shifting our object of attention from questions of environmentalist belief and awareness to more terrestrial worlds of practice and living.

Because the tendentiously universalizing notions of environment, nature, and wilderness were forged and given traction through an entanglement of spiritual and earthly vocation, I am both unsurprised by recent turns to religion in political theory and taken aback by their tenor. In particular I am thinking of the turn by a number of writers to evangelical Christianity for a way to ground a revival of theoretical faith in political universals. The philosopher Alain Badiou, for instance, invokes the Christian figure of Paul as an exemplar for political thought, drawing an analogy between, on the one hand, Paul's fidelity to Christ's Resurrection and, on the other, a militant's fidelity to a political truth.[11] That unflinching fidelity, he argues, is the core of political universalism.[12]

Badiou continues, erecting a conceptual framework of universalism through universal address based on Paul's techniques of argument. But even as I read in admiration I find myself wondering, caught by quotidian questions of vocation that the philosopher does not ask. What happened to Paul, I want to know, that he would change his name and his life? What drew him to persist in witnessing in unfriendly places? I am puzzled, in other words, not by the turn to an example of religious commitment for a way of thinking through political commitment, but by how thinly that comparative example is treated. Paul's writings are mined by Badiou and others for their logics; they become a formal model for universalizing new foundations, declaring common cause, and reimagining what common cause means. But what are causes and logics, for claims of whatever scale, without those subjects who not only feel that things could be ordered differently, but who act to realize another order? How do such subjects come to be? How do they come to act in fidelity to something other than the world they live in? And what other lives of fidelity might be lived or set up as exemplary, aside from that of the unflinching evangelist?[13]

These questions echo and bring us back to the questions of environmental calling with which I began. Let us turn again to Rupert, and in time let us meet William. Let me show you their travels, how they witness through the stands they take and how those stands shape their senses of themselves.

They provide dense and dirty figurations of border-crossing commitment. Cosmopolitan and environmentalist subjects in the making, they act for their environmental principles, making cosmopolitan examples of themselves. Making oneself and one's deeds exemplary is part and parcel of the earthly calling.

Rupert

Rupert and I became fast friends in the months after we met on the Zodiac. I soon asked for permission to study him and his anti-incineration campaign, and in addition to the many months we spent together doing Greenpeace work and research, we often went out or ate together, along with his wife, May, and my partner, Zamira. We eventually did go rock climbing at Shek O, as we had discussed that sunny afternoon on the rock, and we took other excursions together on the Zodiac when the weather was right.

About half a year into my research, Rupert agreed to a formal interview. As the day approached, though, he suggested a change of plans. Why not turn our interview into an outing, one that would include May and Zamira? We would meet at the base of the Peak Tram, a popular tourist attraction, and ride up to the top of Victoria Peak.

That day at the station, Zamira and I easily recognized our friends from a distance. They made a striking couple: Rupert was wearing an unbuttoned long-sleeve shirt over his Greenpeace T-shirt, baggy cargo pants, and his favorite pair of Birkenstocks. May had on a Greenpeace shirt from a different campaign, a light jacket, and sport sandals. Both wore colorful woven shoulder bags. After riding the tram up and drifting through some of the souvenir stores at the top, we settled down with ice cream in a Häagen-Dazs shop for the official interview. It began as a dialogue between Rupert and me but quickly turned into a four-way conversation. Zamira and May both jumped in frequently—Zamira posing follow-up questions, May adding bits of history or analysis as she saw fit.

Rupert joined Greenpeace in Vancouver, working his way up from the organization's bottom rungs. He had moved to Vancouver from Hong Kong with his parents and May in 1993. "June 4"—Rupert, like most people in Hong Kong, uses the common month-day shorthand implying that we all remember the year, 1989, of the Tiananmen Square massacre in Beijing— had deeply shocked his mother and father. The event conjured memories of the Cultural Revolution that they had narrowly avoided by moving to Hong

Kong. So, with China's impending resumption of political sovereignty over Hong Kong, Rupert's family was eager to leave the territory. May moved with them, and she officially joined the family through marriage a year later.

Times were not easy in Vancouver. At first, Rupert worked for a family friend, selling computers, but he grew dissatisfied and unhappy. So, with May's support, Rupert quit his job and began to look for other employment.

This was how he met Greenpeace. "I was a little bit puzzled at the time," Rupert remembered. "We're new to Canada, now I'm jobless, May's not working . . . But it turned out to be all right, because without major change there's no improvement. . . . It was at that time that I opened up the classifieds and saw that Greenpeace was looking for a fund-raiser. But I didn't know the organization at all then; I only knew the Greenpeace images—the whaling, antinuclear campaigns. I didn't know anything about the organization at all."

How little Rupert knew became painfully apparent at his job interview. "I was totally out of place then, because when I went in for the interview, I still had the typical Hong Kong mind-set." May chuckled in anticipation as Rupert dramatically listed the features that betrayed him. "I was wearing a tie, a suit, carrying a briefcase . . . walking into a hippie Greenpeace office!"

Rupert played it up as Zamira, May, and I laughed out loud. "People looked at me like, Hey, what is *he* doing here?" And May joined in, drawing her face into a stern rendition of someone from Greenpeace watching a strange man from Hong Kong walk in. "No solicitors!"

Despite his fashion gaffe, Rupert got the job and started working as a canvasser for the organization. He told me once that the canvassing work introduced him to Vancouver. Dropped off daily by a van in different neighborhoods to solicit new members and donations door-to-door, Rupert walked through parts of the city most of his fellow Hong Kong immigrants never explored—and he gradually learned how to read house exteriors, particularly front landscaping and the cars in driveways, to the point that he could reliably anticipate how he would be received before he rang the doorbell.

"From then on, I don't like wearing suits, shirts, and ties anymore." Rupert looked thoughtful for a moment. Then, with a grin, he pushed the sartorial theme further, "It is then that I started to wear Gore-Tex pants and Gore-Tex jackets!" We all laughed again. It was impossible not to notice Rupert's fondness for high-performance, waterproof, breathable outerwear. "I sort of carried that from Vancouver back to Hong Kong . . ."

May interrupted him here. He always hated suits, she reminded him, whether before or after Greenpeace—even when he worked at a hotel, he didn't like wearing suits.

"Yeah, true," continued Rupert. "Even when I was working for a hotel, even though I had my own personal suit, not a uniform—when I went to work, I still had to wear a T-shirt and shorts. Get to the office, change into this monkey suit, work . . . and then change into shorts and T-shirt again. I never developed a love for suits, shirts, and ties."

The "Hong Kong mind-set" days are over. They end in Rupert's telling through an accidental encounter. Searching for a job, he sees an ad, and he applies even though he knows nothing about the organization. While serendipitous, the encounter is transformative. Rupert represents his altered states rhetorically with clothing; a suit, shirt, and tie stand in for sternness, formality, and stiffness—characteristics of a mind-set that becomes meaningful only when juxtaposed with a "hippie" Greenpeace that renders them culturally conspicuous. Rupert's statement, *From then on, I don't like wearing suits, shirts, and ties anymore*, positions him at a distance from the symbols of that mind-set; it marks him, the suit-eschewing subject, as beyond its reach. Rupert establishes this distance firmly through the use of temporal markers, *from then on* and *anymore*. Using these indicators of time gone by, Rupert puts Hong Kong rhetorically in his past, separated from the present by the watershed event of his joining Greenpeace. Perhaps the greatest evidence of change he offers is his ability to look back and laugh in the context of our interview.

His story of change takes the form of a travel encounter narrative. A man, whose ties to the East are betrayed by his clothing and accoutrements, encounters a foreign way of being over the course of his travels in the West and gradually comes to call the foreign way his own. The mind-set that Rupert claims to leave behind is defined through its imagined location—he calls it a Hong Kong mind-set. His present consciousness, it would follow, must be beyond Hong Kong, and he achieves this "beyond" through his travel experience.

Greenpeace and Vancouver are credited for changing Rupert, but notice too how May's intervention ties Rupert's present life as a Greenpeace activist to a way of being that preceded his environmental work. Whether *before or after Greenpeace*, he hated suits, she urges Rupert to remember in his public statement to me. If suits stand for Hong Kong, then, at least in this recorded history, Rupert has never quite fit in here. Greenpeace's rhetorical function in May's and Rupert's account, then, is not simply to indicate a moment of transformation, but also to represent the culmination of a movement away from typical Hong Kongness that has been long-standing.

The particular icons of Easternness in Rupert's tale—suit, briefcase, shirt, and tie—might in other versions of East/West encounter narratives represent the West; yet here, Hong Kong's particular history as a colonial and capitalist metropolis allows what in Cantonese are called *saijong*, or Western suits, to stand in for a Hong Kong mind-set posed against the Canadian. Rupert will be changed forever, marked by time spent abroad. Even when he resides in Hong Kong, he carries Vancouver back with him in the pockets of his Gore-Tex pants.

• • •

Dinners at Rupert's and May's place are fun affairs. May cooks fabulously, and being in their home is a pleasure. They live just off Pokfulam Road on Hong Kong Island, in a flat Rupert's father acquired when he still worked for the government. With Rupert's parents in Vancouver, Rupert and May now have the run of the spacious, three-bedroom apartment. A piece of dyed fabric hangs in lieu of a door for their bedroom, and the bedrooms are painted in bright reds and greens. The guestroom, which they generally leave empty for when Rupert's parents come into town, looks like they gave up painting before they finished. Rupert has bolted climbing holds into the hallway leading from the living and dining room area to the bedrooms, and his climbing and kayaking gear is stowed by the washing machine in the back room originally designed to accommodate a live-in maid.[14]

Some evenings we watch a movie on VCD. We lounge on futons, and Rupert or May might light some incense to deter the mosquitoes that fly through the open balcony doors. I don't know how the others fare, but I slap and scratch with regularity, wishing we could close the doors and turn on the air conditioner. The balcony overlooks a lush hillside, and when I stand outside and look down, I see no traces of the fruit and vegetable scraps that Rupert and May happily throw off the balcony after dinner in the name of composting.

Other evenings we just sit and talk. Rupert and May tell us many stories about Canada. It was a Canadian friend, someone from Greenpeace, who taught Rupert how to kayak and rock climb. This same friend enlisted Rupert and May in the project of building a hut in the forest where he lived for a while. Another Canadian friend refuses on principle to use soap. Their old boss, Sarah, the one who came from Vancouver to Hong Kong to start the Greenpeace China office, drank beers with everyone after work. When she came to Hong Kong, she would dirty her hands in garbage raids even though she was the executive director. This is so different, they agree, from what a Hong Kong person would do.

Rupert and May still have an apartment in Vancouver, in a bohemian neighborhood near the University of British Columbia. They hope to move back someday. It's not in the same neighborhood where most Hong Kong immigrants live, they tell us. Most of them live in Richmond, a "new China-town" suburb now famous for great food and notorious for ostentatious homes built by some of the wealthier immigrants.[15] You can spend the whole day there without speaking English, says May. Their apartment in Vancouver is a short drive from Richmond, so they can get a taste of Hong Kong when they're homesick or hungry. But, they add, they don't surround themselves with it all the time.

Home works here as a refuge from the Hong Kong mind-set. In it Rupert and May can be environmental citizens of the world, living gently upon the earth. This earth, though, is a specifically Canadian one. It is to Vancouver that Rupert and May long to return. As I survey the signs of environmentalism in their flat on Pokfulam Road—the gear, the textile art, the ecology texts—they seem to me laden with Canadian meanings.[16] The kayak, the climbing holds, the willingness to suffer mosquito attacks—these expressions of an environmental aesthetic gesture to a specifically Canadian notion and appreciation of a wild outdoors. Not only do they mark and enable environmental sentiment, they conjure Canada. Even while in Canada, Rupert and May differentiated themselves from other Hong Kong Chinese by treating the space most occupied by Chinese food and language as a space for visiting but not for living.

• • •

We had gotten sidetracked in Häagen-Dazs. Rupert and May were regaling us with details of a side-trip they took to Montreal on their way from British Columbia to Ontario for a Greenpeace assignment. "Great, great food! Beautiful city!" they gushed. It was fairly expensive, they admitted, "but, well, we're on vacation!" They happened to go during the Montreal jazz festival, and May and Rupert glowed while reminiscing about different performances and venues. It sounded to me, I said, like they had gotten a good taste of Canadian life.

Rupert nodded in agreement, then added, "And most of it is through the Greenpeace angle." He explained it as a matter of time. "I was first in Canada for a very short while: I was in the commercial company for about three months. So I saw Canada as a commercial person for three, four months. Then I saw Canada with Greenpeace eyes, for two *years*. So I guess my impression of Canada would be very different from how a usual Hong Kong person—who migrated to Canada and worked in a bank, for example—would see Canada." He continued to draw the contrast between the great experiences he and May had enjoyed and the attitudes of someone more "usual." "I've met a lot of friends who say Canada is a very boring place—not a lot of life. You know, they went to Canada, they got the passport, they moved back to Hong Kong. For us it's the exact opposite! We prefer the Canadian life to the Hong Kong life so much that, you know, moving anywhere else seemed a little bit strange at that time. Because we hadn't had enough . . . enough Canada."

Notice how Greenpeace mediates Canada. This mediation not only enables May's and Rupert's environmental transformations, it grants them a Canadian experience distinct from the one that others from Hong Kong would have. When Rupert describes seeing Canada with "Greenpeace eyes," he sets himself and May off from other Hong Kong people rhetorically, much as their decision to live near UBC rather than in Richmond sets them off spatially. Both acts convey that while their movement to Canada was part of a much larger movement of Hong Kong people—it found them in the company of what some analysts termed a post-Tiananmen tidal wave of emigration—Rupert and May are different from the "usual" Hong Kong immigrant.[17] "*For us it's the exact opposite!*" Rupert subtly establishes some geographic equations: if commercial mediations lead one back to Hong Kong (or Richmond), Greenpeace makes one long for more Canada.

Greenpeace alone, however, could not bring Canada back. I watched Rupert grow increasingly dissatisfied with his work over the fifteen months I stayed in Hong Kong. He found himself having conflicts at the office. Some of his colleagues told me that while Rupert excelled in media events and actions, he put too little thought and research into his campaigns. Rupert told me, on the other hand, and May agreed with him, that his colleagues had lost touch with the spirit of the organization.

Zamira broached this touchy issue at Häagen-Dazs by asking Rupert and May to compare: in their opinion, were the people who worked in Hong Kong's Greenpeace office hired because they were smart or because they really cared about environmental issues? I worried the contrast might sound crude, but Rupert and May found nothing strange about the question. Not the latter, said Rupert. His colleagues had impeccable academic and activist backgrounds, but he wondered about their environmentalist commitment. "I remember when I first saw them," he said, "the environmental awareness wasn't the environmental awareness that I was used to when I was working with my Canadian colleagues."

May weighed in as well, "I don't think they are very concerned about the environment."

Rupert proffered some illustrative examples. "I mean, when I was in Canada, a lot of people are vegans. A lot of people refuse to use plastic bags. They refuse to use shampoo and soap. They refuse to shave. They refuse to buy new clothes. They refuse to use washing powder for their clothing. But here in Hong Kong the attitude is different. The environment is something that's in a textbook, something for their research. Environmentalism, or protecting the environment, has not become a way of life yet. Whereas in Canada, for a lot of people it became a way of life."

Part of this way of life seemed to be about lifestyle, the way of life they made in their painted apartment and Gore-Tex gear. Part of it extended beyond their home and clothing, to a more general sense of conspicuous rebellion. Rupert's answer had been unequivocal when I asked him why he had wanted to work for Greenpeace: "Because it's a fun organization! I saw them on TV, they're crazy! They drive an inflatable, you know, people dump barrels on them, they stay there, you know. It's a very . . ." Rupert had paused, scrunching his face as he searched for a good word, "respectable . . . ?"

"A very *gutsy* company," May offered.

Yes, that was right. "Lots of guts. I think deep down I'm a rebel. I like to rebel." May had laughed affectionately while her husband explained Greenpeace's allure: "Greenpeace people really make me feel they are . . . Yeah, that's the way to go! Fuck the system!"

Fun, crazy, full of guts—Greenpeace resonated with what Rupert called a rebel deep inside himself. While Wing Hung, whom Rupert called "smart" but not necessarily tuned to the environmentalist way of life, brought specifically anticapitalist concerns to his work in the organization, Rupert's passions were stirred by a more generalized sense of activity and transgression, piqued by charismatic demonstrations captured on television.

Rupert's excitement for big actions was palpable and contagious. One day I found myself dressed in a white jumpsuit, tied-in with a climbing harness and rope on the rooftop of a government building, lowering roller brushes and buckets of red paint to Rupert and Wendy, a feminist comic artist, while they climbed down the side of the building. In fifteen minutes, they painted a four-story-tall welcome message to the developers responsible for the eviction of Wendy and other artists from the government building. This was not a Greenpeace action, but Rupert and Wendy were friends—the arts and activism community was small—and Wendy often drew posters for Greenpeace. Rupert had been eager to return the favor. It seemed like a good idea when Rupert proposed it, I remember thinking to myself as police vans pulled in below and journalists snapped pictures. I came to appreciate Rupert's incredible way with the local authorities that day; after reaching the ground, he captivated the police officers with a smiling show-and-tell of the climbing gear we had used. The police noted his and Wendy's Hong Kong identification card numbers, but they left me alone after he and Wendy insisted that I had done nothing but provide technical and safety support.[18]

Rupert was arrested in Tokyo when he assisted the Greenpeace office there with an anti-incineration banner action. He and three other Greenpeace climbers hung a protest banner on the tallest incinerator in the city. Rupert's eyes lit up whenever he recounted his tale of being arrested and jailed. Our friends and I couldn't help but notice, somewhat ambivalently, the eagerness with which he embraced his new notoriety when he returned to Hong Kong from Japan. That same light registered Rupert's excitement when Greenpeace's flagship, the *Rainbow Warrior*, came to Hong Kong to support his anti-incineration campaign.

Rupert's passion for actions sometimes calls to mind an observation that eco-critic Noel Sturgeon made about another direct-action environmental group, Earth First! "The forms of direct action pioneered or preferred by Earth First! . . . were all tinged with a patina of toughness, risk taking, and military-like stealth that lent itself easily to machismo." Sturgeon points out that a popular self-portrayal of Earth First!ers, an image of "a hairy man" in "size 11 mountain-climbin', woods-hikin', desert-walkin', butt-kickin', rock-n-rollin' waffle stompers," ironically theatricalizes a "redneck" machismo.[19] This machismo is drawn in contrast to middle-to-upper-class liberal environmentalism, making the latter look soft, clean, passive, smooth, and hairless in comparison. The figure of virile activism frequently associated with Greenpeace's stunning and often dangerous actions was perhaps more overtly middle class—with its reliance on technical climbing gear and Greenpeace's reliance on research scientists—but it seemed to me no less macho, as when May or other women like Maria adopted the eco-warrior image to reject stereotypes of passivity.

Rupert's acts of outdoor prowess were gendering as well. The hobbies he picked up in Canada and associated with the Greenpeace way of life—rock climbing, Zodiac piloting—were all read by others in Hong Kong as marks of a kind of foreign masculinity. When Faanshu, for instance, teasingly cried out, "*Wah, hou man, ah!*" as he saw Rupert pull his way up the rock face on our boat trip, he not only poked fun at, but recognized and reiterated Rupert's prowess as excessive masculinity.[20]

Of course, this cosmopolitan masculinity existed only in and through a network of relations—with other figurations of gender, as well as gendered figurations of place. Rupert's primary point of reference and departure was what he identified as "the Hong Kong mind-set." Rupert and May drew a contrast between a Canadian eco-warrior-ness and Hong Kong textbook learning about the environment. "Environmental awareness" in Hong Kong was not the same as that of their Canadian colleagues. The contrast was drawn most strongly through examples of refusal; refusals of consumerism and standard cleanliness indicate Canadian environmentalism as a way of life. These refusals were cousins of the system-fucking rebellion that Rupert clearly relished. Taken together, they delineated for Rupert and May a subject position outside of Hong Kong life—one where Hong Kong locality was imagined to be transcended through environmental activism and lifestyle.

William

"Hi, I'm William Lee." The man spoke English fluently, but I couldn't place his accent. He appeared to be in his mid to late thirties—but his face was a bit boyish, so it was hard to say for sure. His suit was impeccable—sharply cut and dark, it made the white of his crisp shirt pop.

William Lee and I met in 1997 through an introduction by a high school classmate of my mother and father. Like my parents, and like many graduates of Kowloon's Queen Elizabeth School, Uncle Chris had gone overseas to the United States, where he worked for years as an environmental engineer in Connecticut. He had moved back to Hong Kong recently, though, to manage the Hong Kong office of an environmental consulting company. My father, eager to help me with my research, urged me to contact Uncle Chris. I did so, and he directed me to William, a young engineer and project manager in his office who he thought might welcome an outside observer.

William was eager to talk to me and excited to help with my research. Though his training was in civil engineering, he told me, he had developed a keen interest in "the sociocultural," so he was happy to help my ethnographic project in any way he could. A year later, when I returned to Hong Kong, I worked for three months as a volunteer on William's project. Most of my days there I helped him with tasks such as editing environmental impact assessments (EIAS) and mocking up work timelines for bids and proposals. William and the other engineers welcomed the free assistance. One of the company's major studies was already over budget, so every bit of work they did for that job—and there was a great deal of it, since their client, the Hong Kong government, kept demanding more changes to the final document—came out of their own pocket.

William managed me attentively. He took care to ensure that my assignments were interesting and important, and he always introduced me to other consultants as someone with a degree in environmental sciences so that they would respect me. William also came to my defense one day when another engineer confronted me with a Greenpeace pamphlet that featured a photograph of me holding an anti-incineration banner in a protest; William quickly explained that my research was ethnographic and required that I participate on all sides of environmental controversy.

Most days William and I ate lunch together. We took the elevator down from his office into the shopping complex below for takeout—usually Chinese barbecue, Vietnamese noodles, or sandwiches—and brought our meals back upstairs to eat in the office conference room with the other engineers. I loved listening to the stories they told at lunchtime. One of my favorites concerned a job in Taiwan, where the consultants painted a smokestack bright blue; unlike Americans, who would rather smokestacks be camouflaged, residents there wanted the chimney to be made more attractive through color.[21]

These stories, and the chats I had with William and other engineers while we leaned on each other's cubicle walls, were my favorite part of work; they often illuminated external or structural factors that shaped the EIAS I was working on, but they presented me with a dilemma. Could I write about these conversations? Everyone knew why I was there, but as most ethnographers know, lips loosen over time, and audio and video recorders fade from consciousness. So one day William and I staged a more formal interview to be explicit about the fact that we were on the record. For our interview, we opted out of our casual lunch routine and sat down instead for a more sedate meal at an upscale Shanghainese restaurant in the shopping center below the office. Placing my minidisc recorder on the table between us, I began by asking William for basic information about his schooling, upbringing, and places of residence.

"Right, bio-data," he agreed. He started by telling me that he attended six years of primary school and three years of secondary school in Hong Kong, then was sent to boarding school in rural England at the age of fourteen. This luxury, as he described it, he attributed to his father's work for the Hong Kong government in the Lands Department, and he suggested it gave him a unique perspective. "Yeah, I'm actually more . . . Westernized, or Englandized, because I left at fourteen. Then I came back to Hong Kong to work after I graduated. So I actually lived in England for seven years—and, well, probably picked up more . . . more, you know, this British way of thinking." William stayed on in England to complete his bachelor's degree in civil engineering. Afterward, at the age of twenty-one, he returned to Hong Kong in the mid-1980s, securing one of the thirty civil engineering posts available in the Hong Kong government. In 1991 he moved to Australia, where he earned a master's degree in waste management. It was in Australia that he made the shift into the field of environmental engineering, returning again to Hong Kong in 1996 to work as an environmental consultant for a multinational firm.

One of my clearest memories of William is that he loved environmental consulting. He called it "eye opening." When we first met, he had emphasized to me that it was his work as an environmental engineer that had fostered his keen interest in "the sociocultural." In Australia William learned to think about development projects as more than engineering problems. He told me, for instance, that he learned the importance of confronting cultural issues because of the strength of the Aboriginal movement in Australia in matters concerning land use. Another time, he came to my cubicle with an article on Hong Kong archaeology: "Look at this, these ruins are over four thousand years old!" The archaeological site discussed in the article was located near an infrastructural development being proposed by the Hong Kong government; it would need to be accounted for in the EIA. "I never would have read archaeology articles like this in pure civil," William pointed out. In pure civil engineering, you didn't think about issues like archaeology, culture, and sustainability. But now, he exclaimed as he waved a copy of the assessment he was overseeing, engineers needed to use a multicriteria analysis, a global approach. If Rupert's eyes were opened in Canada to outdoor sports, activism, and environmental lifestyle choices, William discovered new ways of thinking about engineering in Australia.

All this was on my mind when I asked William to remind me when he went to Australia.

"1991. Well, it was the Tiananmen Square thing . . . that triggered . . ."

"Yeah, I bet."

"'89, right?"

Meanwhile our rice noodles with shredded chicken and sesame paste had arrived. William rolled up his sleeves and motioned to me to try them. He nabbed a noodle with his chopsticks and paused, rethinking now whether post-Tiananmen fears were a sufficient explanation for his move.

"I did apply for Australia in '89, but for me, I wouldn't think there was a trigger. When I did go, it was more for different things—not really to avoid communism, or the persecution. It had a tremendous impact, but to me, personally, it wasn't strong enough to drive me away. Because while you saw it on TV, it wasn't happening right on the streets. I think it was the chance . . . at the time, Australia opened up for immigration. I was young, we had just gotten married, and I had lived in England before, so I didn't think it'd be a big problem."

For someone else, who hadn't spent time in England, it might have been a bigger problem. "I believed my English was good enough to settle, get a job and everything. That's the reason why a lot of local guys didn't go. They didn't think they would survive in an English-speaking environment. It took me six months to get my first job, but that had more to do with the economic situation at the time. But there are other guys, other Hong Kong graduates who spent two years there and never got a job. When employers looked at your CV they'd say, Oh, all right, this guy got a degree in England or Canada. Right. So he has more competence in English . . .

"Anyway, I find even interacting with the graduates in Hong Kong—who are, of course, the elite . . . who at Hong Kong U would be like the cream of Hong Kong's civil engineering—. . . when I interact with those guys, you know, they're bright guys, but because I had a very different secondary education, and I've observed the same thing is true for all the graduates in

Canada, the UK, or wherever. We're actually more vocal, I think more natural thinking still.

"Those local graduates, they're bright, but they've been trained in a different system. So once they go into government, they want to be government. . . .

"See, the test came when the immigration wave came. None of those guys actually quit government. All the guys who got in, the Hong Kong U graduates, they're now all still in government. But from the overseas side, a few people emigrated, some quit government and joined private sector. Probably because of the way they grew up, they don't necessarily see government as their preferred career—you know, the 'real world.' I think the test is now coming for those guys who are in government now, because the public sector is now getting major pressure for reform. And that would mean, they wouldn't have a very good prospect now, because they will be reformed, they will be cutting places."

• • •

William's story, like Rupert's, is of an overseas encounter with environmentalism. The "eye-opening" experience of becoming an environmental engineer takes place in Australia; it introduces him to new ways of conceiving development. Like Rupert, William uses the interview as an occasion to tell a story of how a life abroad, in both England and Australia, has transformed him. The words he uses to describe this change—*Westernized*, *Englandized*—attribute power to geographically imagined regions. This is a wide-ranging power; it not only imbues him with a *British way of thinking*, it also exposes him to analyses of sustainability and forces him to reckon with Aboriginal claims to land.

Notably, William makes explicit, while Rupert does not, his class privilege. William acknowledges the luxury of attending school overseas. William's father, as an official in Hong Kong's colonial government, would have been generously compensated for his work, perhaps receiving a stipend for housing and his children's education in addition to an ample salary. Education in England would have paved the way for William to become, like his father before him, a member of the elite within the colony.

William sets himself apart from *other Hong Kong graduates* and *a classic Hong Kong graduate*. Initially, it seems the main distinction is English education and strong English skills. We soon see that the difference runs deeper. The different educational system in England, William suggests, actually yields essentially different kinds of people. Hong Kong graduates, for their part, become fleshed out in contradistinction to William and his cohort of overseas-educated graduates. *We're actually more vocal, I think more natural thinking.*

In William's narrative there emerge two arenas in which he stands apart from the typical elite Hong Konger. One is private industry, *the real world.* Few Hong Kong graduates, says William, enter the private sector because they favor stable jobs in government. The litmus test proving this difference is the post-Tiananmen immigration tidal wave of the early 1990s. By staying put, the classic Hong Kong graduates fail the test.

The second is employability, as pressures for a leaner government in Hong Kong lead to the cutting of government jobs. This, William suggests, will reveal the inability of Hong Kong graduates to respond. *They will be reformed.*

· · ·

"Then I diversified into environmental [engineering]," William continued, "because of Australia's picking up the environmental movement, in the early '90s. A lot of water management drainage work. Or coal mines, right? Australian authorities were tightening up on the environmental licensing in the coal industry, so there was tremendous growth in the industry. I was lucky enough to get in. Changed jobs a couple of times."

"All in the private sector over there?"

"No, I had a six-month stint with the local government. As a civil engineer. My first employer had a downturn, so I got retrained for a year with them, and then I worked for government for six months. And then moved back into consulting again. Much bigger repertoire coming in."

William chewed a mouthful of Shanghai-style *dan dan* noodles and considered the obstacles he had surmounted over the course of his career. These were not experiences, he reiterated, that other Hong Kongers would be likely to have. "Hong Kong people wouldn't want to have this. Six months get a job, work there for a year, then you got retrained. . . . I had to really . . . psychologically gear and say, 'Hey, I'm good enough! I need to get back in!' And I did. Hopefully I improved myself through that experience."

William now explained the context for his lesson. "Australia was going through economic reform, after the bubble burst of 1987 . . . In 1991, when I arrived, it was still restructuring. My first experience had been in government—a secure environment. But in Australia, I realized I had to earn a living! I had to be good at what I was doing! You're constantly on the lookout. Improve.

"Which, incidentally, is not unique to me. If you saw the local Australian guys, white guys, they had the same problem. They had a good time for so long, then they realized, 'All right, I'm losing my job.'" William thought about the equivalence he set up, then reconsidered. "Probably even a more severe impact on me, as a newcomer."

• • •

If William's first proving ground was the private sector, that is, his decision to leave the safety of civil service to pursue work in "the real world," his second was Australia itself. Australia becomes in William's narrative a kind of sorting ground, a real-world place where *a lot of local guys didn't go* because they *didn't think they would survive.* Here, William not only establishes himself as atypical simply by daring to go there; he proves himself further by succeeding.

Gendering and locating work are as thick here as they were for Rupert, but they take a different form in William's narrative. Rather than citing lifestyle and rebellion as marks of difference, William invokes professional competence and bravery. He assimilates his different experiences into an oral resume. His brief stint in government, a spate of job-hopping and retraining in the midst of an economic downturn—these become an amplified repertoire. William then sets up this repertoire against the experiences of the Hong Kong *local guys. Hong Kong people,* he claims, *wouldn't want to have this*— the instability, the frequent changes of employment. Over the course of the interview, William implies they would be unable psychologically to handle his experiences.

In the end, the Hong Kong *local guy* that emerges in the course of William's and my conversation is a college graduate, security seeking, with little tolerance for economic instability and lacking confidence and English skills that might help him compete in an international job market. So when William recounts how he *psychologically geared* and *improved* himself and got back in the job market, he rhetorically proves and produces himself beyond the capacities of the Hong Kong man. He becomes a kind of neoliberal ideal— adaptive, risk-tolerant, globally mobile, responsible for his own competitiveness for employment.

In the process of doing so, William also establishes, only to dismantle later, an equivalence between himself and white Australian men. He sets up the equivalence by saying that the obstacles he faced and overcame are the same ones that confronted *local Australian guys, white guys*. His success not only distinguishes him from Hong Kong men but also certifies his passage into Australian managerial masculinity. A breath later, however, William quickly sets himself apart from Australians as well. His statement, *Probably even a more severe impact on me, as a newcomer*, recognizes a cultural milieu and market that perceived him as foreign, and alludes to stories he told me on other occasions about problems with xenophobia and racism. We are left to complete William's unfinished point: if the local economic instability had an *even more severe* impact on him than on the local Australian guys—and he persevered and succeeded—then his story of survival and accomplishment hints at a competence and ability to weather professional danger and risk that is even more potent than the locals'.

As William moves through his account, he builds a constellation of relevant actors, alongside which his life history emerges and acquires meaning. The *other* guys, the Hong Kong guys. A term like "other" implies difference, to be sure, but it can only do so by also asserting comparability. The differentiation of William from "other guys" is in fact simultaneous with, and is what conditions, the figuration of William as a guy.

. . .

William not only talked about his difference from the other Hong Kong guys; he manifested it through his work. For instance, he once tried to write up a proposal the way he would have in Australia, with a public consultation. "The way the Hong Kong government runs its process is still top down," William explained. "In the Australian system, you have a consultation meeting early on in the process. Although the government advocates certain things, the process actually allows more input up front."

"In Vietnam when I did a World Bank job, the Bank mandated a minimum of two consultation meetings. But in Vietnam, we couldn't do what we wanted to do because the local people say, Hey, you can't talk with these people, you have to go through the political system."

"Hong Kong's actually just as bad. There's no consultation. You do a draft report, you submit it to the ACE, the Advisory Council on the Environment, which is government appointed, and it's not until the last phase of the process—where all your design, everything, has been determined—that you do this report, go to the Parliament or LegCo [Legislative Council]. By the time you go to LegCo, you've burned your funding, right?"

It irked William that "consultations," the forums where public opinion is supposed to be solicited so that local concerns can be addressed in a project's design, take place so late in the process. When they happen late, it ceases to be feasible financially for a client or consultant to address any concerns that are raised. Furthermore, William suggested, the consultations seemed designed to transmit a particular message about a project's benefits, rather than to hear concerns and questions from the public.

"It's actually a public *information*, not a consultation!" He sounded adamant now. "I'm telling you, I've been there. I tried in my proposal, the way I did in Australia. I stuck all those workshop meetings up front in the project . . ."

"What happened?"

William paraphrased his client's response: "Don't do it this way."

"Wow."

"See, in Hong Kong, it's all been determined. So consultants do technical work; that's all they want. They want you to front the public presentations. Presentations, not consultations. The socioeconomic issues are deemed to be taken care of in the planning process, the framework. The EIA in the World Bank has scientific studies, sociocultural, and economics. But in Hong Kong, it's only biophysical."

I recalled then that even in the study I was helping William prepare, the sociocultural section was small. In a document of several hundred pages, the ten-page section looked meager. The methodology for determining the sociocultural impacts had consisted mostly of a walk through the area. And while the discussion mentioned that people lived nearby, the authors had spoken with none of the residents.

"I put that in there," emphasized William. He could have not included it, and it would have been fine. "Look around, not many people do it. I put it in there because the project's such a difficult topic anyway." The project was unpopular in the public eye. "But there is no consultation; they don't want us to talk to anybody. They keep it a secret. Though they decline to say, You should not release any information, anything has to come through them."

William criticized a Hong Kong system that was incomplete. Unlike the Australian process to which he was accustomed, the Hong Kong process of determining environmental impact only paid lip service to the issue of public consultation. It had all been determined. He noted indignantly the difference between what he was forced to do in Hong Kong and what World Bank expectations would have been. He indicted the Hong Kong system for deciding that a project is socially acceptable before the EIA is even written.

This was strong talk of an ethical kind. Principled talk about principled action, and as a public act, it was talk that drew a distinction in the world, between William and the Hong Kong system. William tried to do his work in a different manner. *I tried in my proposal, the way I did in Australia. I stuck all those workshop meetings up front in the project.* Despite the *top-down* practice in Hong Kong, he attempted to create a forum in which the public could discuss the implications of his project, and he identified this attempt as one of trying to do things as he did them in Australia. Later, though few other companies address more than the "biophysical" in their study documents, William made sure to include a section on "sociocultural impacts" and proudly claimed ownership of it: *I put that in there. Look around, not many people do it. I put it in there.*

William's words and actions indicate an ethics where technical practices and methods for assessing environmental impact have a meaning beyond their function as planning techniques. Specific methods connote locales and subjects. William's decisions to make space for discussing sociocultural impacts in both the process of planning and the final study document, even when his client does not necessarily care to consider them, reflect a sense of professionalism — a commitment to professional standards located beyond Hong Kong, in Australia.

Comparisons

I close by repeating Rupert's words because they illustrate so well the extent to which he and May understood Canadian hippieness as a normative environmentalism, one beyond the reach of his Hong Kong colleagues yet framed as a telos through his hopeful phrasing "not . . . yet": "I mean, when I was in Canada, a lot of people are vegans. A lot of people refuse to use plastic bags. They refuse to use shampoo and soap. They refuse to shave. They refuse to buy new clothes. They refuse to use washing powder for their clothing. But here in Hong Kong the attitude is different. The environment is something that's in a textbook, something for their research. Environmentalism, or protecting the environment, has not become a way of life yet. Whereas in Canada, for a lot of people it became a way of life." Rupert's comparison, with its mild, perhaps naive, condemnation of Hong Kong environmentalists, illuminates some of the blind spots of North American notions of appropriate appreciation for environments and wilderness. And it makes obvious the cultural specificity of environmental aesthetics and ethics; Rupert consistently identifies his own sense and judgment of what makes for an environmentalist way of life via his trajectory through Canada. This is an act of self-reflection, and it is a recognition of the situatedness of what constitutes environmentalist practice that is sorely missing in the actions of most Northern environmentalists. I appreciated that Rupert was willing to name the lure of rebellion, of big action, of fucking the system. It makes obvious the gendered and gendering appeal of his environmental vocation. To be a rainbow warrior is to perform a particular oppositional militancy that feminizes and devalorizes what it stands against, including the suits and ties of tycoons and businessmen, the supposed parochialism of people in Hong Kong and Canada, and the softness of urban and intellectual life. Only slightly less obvious is the appeal William finds in his talk of constantly surpassing risk-averse, less vocal, Hong Kong guys seeking secure civil service jobs.

For Rupert and William, the environmental marks a space of transcendence—a transcendence of prior ways of thinking and a transcendence of the local. It not only represents an arena of new ideas and activities but also associates them metonymically with another place. Environmental transformation is located outside of Hong Kong in their narratives, and Rupert and William produce an equation of environmental practice—environmental activism or environmental engineering—with a notion of hailing from somewhere else. These are practices of self-care that do not exactly transport them but thicken their attachments. Wound up in their earthly vocations are ethics of geographic difference. Rupert and William distance themselves from a fictive, typical Hong Kong mind-set, and they do so not simply through travel stories. They realize that distance every day through environmental practices themselves.

From one perspective, we can see environmentalism as an arena in which locating, translocating, and gendering modes of life are worked out and articulated, where the workings out in turn constitute that arena. Many of these modes of living oppose dominant cultural norms in Hong Kong, such as the norms exemplified by the merchant mandarin that Aihwa Ong identifies in her study of the cultural logics of overseas Chinese capitalism.[22] Rupert's narrative of environmental masculinity, for instance, parochializes the "Hong Kong mind-set." Similarly, Wing Hung's global solidarity with the antiglobalization movement explicitly questions the capitalist accumulation signaled by tycoons like Hong Kong's Li Ka-shing. These alternatives, while liberating in some respects, are also nourished by and perpetuate certain dualisms—for example, those between Hong Kong and the outside, the vocal and the silent, the active and the passive, the real world and the academic, the global and the local.

From another perspective, however, environmentalism is no mere arena—by which I mean it is more than a context for other politics and interests. Specificities of environmentalist practice—and specifically environmentalist practices—are crucial to the forms of care, both self-care and environmental care, that Rupert and William cultivate and that cultivate them. These are earthly vocations, oriented to environmental protection at the same moment they are shot through with other significances and investments.

I am tempted to go on—two examples don't feel like enough. I want to compare Rupert and William with Wong Wai King, the woman in Tai O whom we met in chapter 2, who learned about oral histories from visiting researchers, and whose activism to preserve local culture kept her in contact with students, journalists, academics, and others from Hong Kong and beyond. I want to remind you of Wing Hung and his passions. I want to remember Maria with her blonde hair, *gwailou* boyfriends, and gamete-drawing daughter. Together, they not only exemplify how modes of being, feeling, and identifying with worlds outside one's supposed own are (at) the very heart of environmental action in Hong Kong. They point to how their supercession of immediate location enables imagination of, and action for, a political alternative. They are reminders of how differently located and locating such lives can be. They might simply offer more evidence that environmental vocations have their pleasures. Distinction itself is one such pleasure; fucking the system is another; worldliness is another, as is global solidarity. These various environmentalist cosmopolitanisms have as their enabling conditions certain structuring details of Hong Kong and global history—anxieties brought about by Tiananmen, educational networks and paths of privilege forged through British colonialism, the financial and technical infrastructures built to support Hong Kong's use as a treaty port and financial hub, and existing anti-imperialist movements—even while their very point is to transcend context in their universal address.

Assembled, these examples would surely convey the importance of cultural, historical, geographic, and biographic idiosyncrasies. They are all specific, and yet inescapable for all of them, and for me in writing about them, is the practice of drawing comparisons and distinctions. Earthly vocations hinge on such practices of cosmopolitan and comparative self-care. The activist's investment in activism dovetails with being a different kind of Hong Konger and a longing for Canada. The environmental consultant's investment in writing sociocultural impact assessments and holding public consultations as he would in Australia, even when Hong Kong policies do not require them, is part of his ongoing practice of being different from "the other guys." Principled fidelity here goes hand in hand with self-distinction, the rewards of making one's life—one's home, one's body, one's work, one's self-presentation in an interview—an example of one's difference.

We picked our way through the remains of an old school. Battered, crumbling walls. It is so sad, she said. It was so beautiful before. It was August 1998, a year before I went back to Hong Kong to research in earnest. We were hiking behind the gleaming new town of Tung Chung, whose high-rise apartments beckoned in soft pastels to a promise of more people.

She looked at me with a wide, serene smile. She was a country girl, she had said earlier. Then with a laugh she added, a fake country girl, for she had grown up in the city. She was from the Hong Kong side; she didn't remember when she had last crossed the harbor into Kowloon. She was a student of comparative literature at Hong Kong University.

A fake country girl, but she meant it. Years earlier, she had bought part of a house in Mui Wo, a village on the other side of Lantau Island. The house was three stories tall, as they all are, set back in the hills in a field of ginger flowers near a crop of mustard greens that a woman came from time to time to tend. She had the second level. Many of the flats in such houses come with all the modern conveniences, like her upstairs neighbors'. But she fancied old things. She loved her creaking windows with ornate latches.

As we walked I told her about my interests in what environmentalism looked like and did in Hong Kong. She listened to me patiently, then with her same wide smile she said, to be honest it sounds like a very Western research project.

I sat with that a while, sometimes I still do, bemused in a ruined British fortress, wondering what to say. Wondering what was lost and what was gained in thinking of everything in Hong Kong as either Western or Chinese, or even as a mixture of the two. Wondering what it might look like not to. Then giving up—for these categories and the comparisons they made possible were as unavoidable in Hong Kong as sand flies in Mui Wo.

A week later I went back to Tung Chung, this time with Siuming, Wing Hung, Faanshu, Ginny, and Ling. We met at the entrance of the MTR station. We hiked from there to Tai O, where Siuming's grandmother lived, sweating like pigs in the heavy heat. Oh, what suffering, wailed Ginny all the way up a long hill. A few meters from the top, we stopped so Faanshu could answer his phone. It was his sister—she wanted to know what refrigerator to buy. Then we kept walking. There was a temple to see. And in Tai O, beers, buttered toast, and dried cuttlefish were waiting.

AIR'S SUBSTANTIATIONS

Hong Kong writer Xi Xi opens her experimental short story "Marvels of a Floating City," a mixed-media piece that weaves together brief narratives and reproductions of paintings by René Magritte, with a fantastic image of a metropolis—a thinly veiled Hong Kong—emerging from the sky.

> Many, many years ago, on a fine, clear day, the floating city appeared in the air in full public gaze, hanging like a hydrogen balloon. Above it were the fluctuating layers of clouds, below it the turbulent sea. The floating city hung there, neither sinking nor rising. When a breeze came by, it moved ever so slightly, and then it became absolutely still again.
>
> How did it happen? The only witnesses were the grandparents of our grandparents. It was an incredible and terrifying experience, and they recalled the event with dread; layers of clouds collided overhead, and the sky was filled with lightning and the roar of thunder. On the sea, myriad pirate ships hoisted their skull and crossbones; the sound of cannon fire went on unremittingly. Suddenly, the floating city dropped down from the clouds above and hung in mid air.[1]

I love this image. It transforms a city that can feel dense and overwhelming into a thing of quiet and delicacy. Xi Xi shows Hong Kong as a place moved by the slightest touch of a breeze, as a place that can become absolutely still. It reminds me of the Hong Kong I sometimes encountered on late-night walks past the government buildings, while taking the slow ferry between Hong Kong and Lantau Island, and at times while sitting on MTR subway trains when, following the example of many others around me, I would put on my headphones and take a nap.

Xi Xi's conceit also turns Hong Kong into something like a natural object, something nearly elemental. The city's mercantile and military origins become almost atmospheric, a storm depicted by layers of clouds and a sky filled with flashes and roars. The pirates themselves—the British Lord Palmerston and the others—are absent in this picture (their presence is marked only by the crossed flag that is raised into the sky), but the meteorological impact they had in birthing the floating city is made clear.

Xi Xi's pairing of city and sky is fanciful and metaphoric—the images of dangling and floating recall the questions about an uncertain future that preoccupied Hong Kongers in the late 1990s—but for me, Xi Xi's image is particularly compelling because it also invokes something profoundly literal. Air is central to the understanding and experiencing of Hong Kong.

To explain what I mean by this, I need to tell another story of city and sky, this one just slightly less fantastic. In April 1999 Tung Chee-hwa visited the headquarters of the Walt Disney Corporation in Los Angeles. The visit was perhaps intended as a triumphant exercise of social capital, meant to perform and to buttress a relationship forged through a controversial agreement Tung had signed earlier that year between the Walt Disney Company and the Hong Kong government. The agreement amounted to a joint business venture. Disney would build a theme park in the Special Administrative Region, a park that would not only serve as a draw for international tourists but also (Tung hoped) provide service sector jobs to the increasing—and increasingly vocal—ranks of the unemployed in Hong Kong. In return, the Hong Kong government would be the primary investor. The agreement would be criticized roundly for its environmental oversights as well as for the economically vulnerable position it forced upon Hong Kong. At least in the Walt Disney Company, though, Tung had a supportive ally. They were in agreement: a world-class park for a world-class city was exactly what Hong Kong needed.

Unfortunately, Tung's visit to Los Angeles was marred by more doubt and criticism, this time from Disney itself. Michael Eisner, Disney's chief executive officer, took the opportunity to express concern about the poor air quality in Hong Kong, noting that it did not mesh particularly well with the family image that Disney so prided itself on cultivating. Eisner never said explicitly that Disney's continued participation in the theme park idea hinged on smog reduction. But people with whom I later spoke—shopkeepers, environmental activists, and taxi drivers alike—would interpret the event as more of a threat, as though Eisner had taken Tung aside and whispered in his ear that Disney would pull out if Hong Kong's air quality did not improve.

One could have remarked upon the irony inherent in this moment when a corporation based in, and associated so strongly with, smoggy Los Angeles faulted another city for its poor air, but Tung made no attempt to do so. Instead, he returned to Hong Kong and sheepishly reported the exchange to his advisers and to the Hong Kong public through the news media.

The newspapers had a field day. Hong Kong had just coughed its way through the most polluted winter in its recorded history. Many residents had checked themselves into hospitals citing respiratory problems. The poor air had also forced my partner and me to relocate from our apartment in Sai Ying Pun, an aging urban district in western Hong Kong where we had been living since our arrival, to a flat in a house in Mui Wo, a village on the coast of Lantau Island. Zamira had suffered three sinus infections in six months. It was time to move.

I remember feeling a guilty sense of relief when I read the news. The extremity of the air pollution—the worst in history—made Zamira's illness, and our move from city to village, count as a moment of participation in a genuinely Hong Kong experience. Until then, I had sought to cultivate indifference toward air and air pollution. Although we, like our friends, routinely avoided waiting or walking on busy streets because the air stung our eyes and throats, and though we often left the city on weekends to escape the pollution, I consistently refused to comment upon or even to notice the air. My justification was simple, if not simple-minded: the people I met in my first months in Hong Kong who were most vocally critical of the air quality were almost without exception expatriate businesspeople from the United States. I did not want to be associated with them. The air pressed upon me, for instance, at a cocktail party celebrating the publication of a book by the renowned Hong Kong landscape photographer Edward Stokes. I was chatting with a representative from the American Chamber of Commerce and his wife when it happened. Hong Kong has to see, she told me, that the environment is an economic problem. Hong Kong wanted to build this Cyberport, for instance, but who would want to come to Hong Kong to work if the air was so bad? If you could not even see? This was the first time, but certainly not the last, that I heard Hong Kong's air coupled with the future of its economy.

At the same time, many of my Cantonese-speaking, Hong Kong–born friends often vocalized their suspicions that politicians who built campaign platforms on the topic of air pollution were motivated by selfish and middle-class interests. Such politicians were only trying to preserve real estate values for the properties of elites, they said. So, in what I considered an ethnographer's effort to immerse myself in an ethics grounded in Hong Kong's particularity, I tried hard to act as if the air stinging my throat were commonplace, not worthy of notice.

But Zamira's illness, the record-breaking winter pollution, and the Disney debacle together forced me to take notice of the air that had been swirling everywhere around, above, and through me and everybody else the entire time I had been in Hong Kong. I remembered then that during my first field visit to Hong Kong in 1996, when I had asked officials about the pressing environmental issues, air quality was always one of the first to come up. Not only that, but air had mediated ruminations about Hong Kong's impending political transition to Chinese sovereignty. "The real concern is transborder pollution," the official at the EPD told me during an interview months before the handover in 1997. "How will we deal with the air and water pollution that comes down from the mainland?" The air is framed as a threat from the north in these pre-postcolonial months. What remained to be seen, they said, was how the Chinese government would respond to Hong Kong's attempts to reduce air and water pollution in mainland China. We will soon see, they seemed to be telling me, what the implications of the handover will be. One activist told me explicitly that they were trying to lie low, and that rather than making any political demands they would concentrate on building relationships with mainland bureaucrats before the transfer of power.

This account of my gradual awakening to the significance of air mimes a standard trope in ethnography, that of the epiphany in, and of, the field. But it is also something else, or it can be if attention shifts away from my eventual ethnographic realization and focuses more closely on my initial attempts to disavow difficulties with the air. That disavowal was plainly an endeavor to distance myself from expatriates; it was a localizing and nativizing enterprise, one whose motivations were analytically untenable but nonetheless impossible for me to resist. If I avow that at stake in my initial refusals was a naive dream of being a Chinese American anthropologist more able to stomach an everyday, everyman Hong Kong life than my imagined doppelgängers, the well-paid expatriates (including those of Chinese descent), it is only to point out that whatever lines of distinction I imagined—and whatever manners I saw available to identify with some people and to distance myself from others—themselves point to the key issue. Air mattered powerfully in Hong Kong. It mattered in deeply felt, variegated, and variegating ways.

All That Is Air

Air matters too little in social theory. Marx famously described the constant change that he saw characterizing a "bourgeois epoch" as a state in which "all that is solid melts into air," and that provocative phrasing served in turn as a motif for Marshall Berman's diagnosis of "modernity" as a shared condition in which all grand narratives were subject to skeptical scrutiny.[2] Yet aside from signifying a loss of grounding, air is as taken for granted in theory as it is in most of our daily breaths. This is unfortunate, because thinking more about air, not taking it simply as solidity's opposite, might offer some means of thinking about relations and movements between places, people, things, and scales that obviate the usual traps of particularity and universality. These traps themselves, it will turn out, are generated through an unremarked attachment to solidity.

To understand this attachment, it is helpful to revisit the context and afterlife of Marx's commonly cited line. The passage where it appears is about a sweeping change:

> The bourgeoisie cannot exist without constantly revolutionising the instruments of production, and thereby the relations of production, and with them the whole relations of society. Conservation of the old modes of production in unaltered form, was, on the contrary, the first condition of existence for all earlier industrial classes. Constant revolutionising of production, uninterrupted disturbance of all social conditions, everlasting uncertainty and agitation distinguish the bourgeois epoch from all earlier ones. All fixed, fast-frozen relations, with their train of ancient and venerable prejudices and opinions, are swept away, all new-formed ones become antiquated before they can ossify. All that is solid melts into air, all that is holy is profaned, and man is at last compelled to face with sober senses his real conditions of life, and his relations with his kind.[3]

Marx argues here that with capital as such comes a constant revolutionizing of society. This is a liveliness of capital. When surplus value is a motivating abstraction, what once were means to generate differential value—the instruments of production—can become a fetter to that project when those instruments are fixed and ubiquitous. A technology might at one time lower the costs of production or enable new forms of goods and markets, but if that technology becomes ubiquitous in a given market through others securing similar means, the advantage it offered disappears. One might try to revive dead capital through new markets, but if it cannot be resuscitated, something livelier must take its place.

Marx's rendering of this process of endless dynamism hinges on a remarkable figuration of solidity. On the one hand, solidity stands for fixity and reliability. The phrase "all that is solid" renders firm industrial society and the long-standing nature of relations among people and between people and land. On the other hand, this very fixity is itself *historical*. Solidity, in other words, is not fixed at all. Marx materializes this paradox of simultaneous fixity and nonfixity in his language, through his images of relations being "fast-frozen" or "ossified," for these images beg the question of what existed before the freezing and ossification. His images of solidification as a process imply a prehistory, one of pre-solidity.

There are typically two responses to such an image of the world where solidities dissolve. A philosopher might strive for some contingent conceptual fixities to make sense of this swirling about. Marx does precisely this in his analysis, and it requires an unavoidable universality. We hear in the passage a mantric repetition of "all." "All social conditions," "all fixed, fast-frozen relations," "all new-formed [relations]"—together they aggregate, yielding an image of a whole that in turn gives way to the epochal atmospheric world of capital. Similarly, social theorists since Marx have sought to develop general terms, such as "flexible capital," "postmodern condition," and "neoliberalism," to grasp and contain a world of dynamism and change.[4]

Another response, one common among cultural anthropologists today, is to refuse the universalizing gesture and perhaps even the very project of the concept. This might take the form of repudiating either the claim that "everything" is melting or the idea that there can be "whole relations" in the first place. Such abstractions kill, this response goes, doing violence to particular human lives and practices that lie outside the terms of the analysis, and such lives are accessible only through empirical work.

The first response is the one usually charged with being up in the air, with not being concerned with concrete details, particular conditions, specific lives on the ground; but in fact, both responses are of a piece. Both responses, whether universalizing or particularizing, seek solid analytic ground; and both find their ground through resort to a "one." This is so whether the one is the unifying one of the "all," or the irreducible particular one refusing subsumption into the general. The conceptual one and the empirical one are a conjoined pair, and both suffer vertigo without firm footing.

Air is left to drift, meanwhile, neither theorized nor examined, taken simply as solidity's lack. There seems at first to be no reason not to let it. When solidity is unconsciously conflated with substance, when only grounding counts for analysis, air can only be insubstantial. We are stuck with the twinned ones—universal and particular—grounded, fixed, and afraid.

Environmentalists in Hong Kong, however, would press us on this attachment to the ground, as would Marx himself. The environmentalists would ask, Is not this stuff floating above and around us itself deeply substantial? As for Marx, we should remember that his claim is ultimately about a dialectics of solidity. Solidities all have a pre-solid past, and air lies in solidity's future. As he declares in a speech during the anniversary of the *People's Paper*, "The atmosphere in which we live weighs upon everyone with a 20,000 pound force. But do you feel it?"[5] It would be a mistake, in other words, to search only for ground when above and around us is substance aplenty. Our living with this substance, furthermore, is neither universal nor particular. Air is not a one, it does not offer fixity or community, but it is no less substantial. The question is whether we can feel it.

Hong Kong might help us feel it. From a certain point of view, there is no "air" in itself. Air functions instead as a heuristic with which to encompass many atmospheric experiences, among them dust, oxygen, dioxin, smell, particulate matter, visibility, humidity, heat, and various gases. The abstraction of air does not derive from asserting a unit for comparison or a common field within which to arrange specificities, but through an aggregation of materialities irreducible to one another (including breath, humidity, SARS, particulate, and so forth). Thinking about the materiality of air and the densities of our many human entanglements in airy matters also means attending to the solidifying and melting edges between people, regions, and events.

This might help us to imagine a collective condition that is neither particular nor universal—one governed neither by the "all" nor through the "one nation, one government, one code of laws, one national class-interest, one frontier, and one customs-tariff" that Marx envisioned, nor even the "one planet" of mainstream environmental discourse. Instead, it orients us to the many means, practices, experiences, weather events, and economic relations that co-implicate us at different points as "breathers." I like this term, "breathers," which I borrow from environmental economics; it refers to those who accrue the unaccounted-for costs that attend the production and consumption of goods and services, such as the injuries, medical expenses, and changes in climate and ecosystems. I like the term because its very vacuousness constantly begs two crucial questions that are both conceptual and empirical: What are the means of counting costs? And who is not a breather?

• • •

The story of air's substantiation in Hong Kong hinges on acts of condensation, and this chapter engages in parallel acts to condense that story. Consider how the pollution-monitoring stations dotting Hong Kong yield a measurement for respirable suspended particulate. Enclosed machines on rooftops and streets ingest millions of mouthfuls of wind a day, calming it so that the particles it holds can be collected to count, to accumulate enough of the particular for it to register as weight, as substance worth talking about. Miming this method, I collect the details in a diffuse set of contexts: the production of air pollution as a local and global medical concern, the material poetics of *honghei* (air) in daily discourse and practice, the acts of large- and small-scale comparison signaled by air, and the transformations that condense Hong Kong's air into measurable particles and then further into a particular, yet internationally recognized, metric for risk.

In short, four forms of air concern me: (1) air as medical fact, (2) air as bodily engagement, (3) air as a constellation of difference, and (4) air as an index for international comparison. Ultimately, my aim is to gain a deep understanding of all of them and to move seamlessly between their methods and registers. Rather than focusing on just one, I make a start in each of them because conveying the dispersal of air's effects and its substantiations is one of my chief aims. This has produced a text that can seem diffuse; its argument requires some work to condense. But that is exactly what people concerned with air must do: turn the diffuse into something substantive.

Air and Dying

Climatologically, there are two Hong Kongs. Beginning in May and June, the air in Hong Kong swells as winds blow in from the tropical south, bringing heat and humidity. Temperatures will range from the mid-eighties to the high nineties Fahrenheit, while the humidity hovers around 95 percent. The air sticks to you as you walk, forms a sheen on your skin as you move from an air-conditioned bus, taxi, or building to the outside. In the late summer, there are the typhoons—great oceanic whirlwinds that occasionally batter the small island with wind and rain as they spin through the Pacific. In colloquial Cantonese, typhoons are called *da fung*, the beating wind.

Then, around late September, the winds begin to shift. Cooler and drier air gradually blows in from the north, across mainland China and Asia. The temperatures can plunge into the mid-forties—as they did in the winter of 2000, when the streets filled with puffy North Face jackets—while the humidity drops to 70 percent. In these drier months, Hong Kong can feel temperate. In the summer, the air in Hong Kong is heavy with heat and water, but in the winter months its weight comes from a different kind of load as the cool, dry winds sweep the smoke and soot from the skies above China's industrial factory zones into Hong Kong.

It is these sooty winter months that most likely motivated Michael Eisner to pull Tung Chee-hwa aside during Tung's visit to Los Angeles. If Eisner's criticism of Hong Kong's air was indirect and vague, the critiques voiced a few years later by Hong Kong doctors were specific and direct. In 2001 and 2002, faculty from the departments of Community Medicine at the University of Hong Kong (HKU) and the Chinese University of Hong Kong (CUHK) published separate articles in internationally known scientific journals linking Hong Kong's air pollution and declining health. The first of the two, "Effect of Air Pollution on Daily Mortality in Hong Kong," appeared in the journal *Environmental Health Perspectives*. The second, published in *Occupational and Environmental Medicine* by researchers from CUHK's Department of Community and Family Medicine, was titled "Associations between Daily Mortalities from Respiratory and Cardiovascular Diseases and Air Pollution in Hong Kong, China."

The articles' findings were chilling. Both studies concluded that acute air pollution had significant short-term health effects. More people died of cardiovascular or respiratory illness on days with bad air quality than they did on days of good air quality. The HKU study also compared warm- and cool-weather data and found that the chance of pollution-correlated mortality was statistically higher in the cool season.

Both articles take pains to locate themselves in a citational network. I mention this not to argue that citational networks are invoked to confer authority upon the articles, a point well argued by others already.[6] Instead, I am interested in the warp and woof of the network being woven, for it lends a specific character to the objects and political substances emergent in it.

One way to see how is through the titles of some of the citations that form the network:

"Particulate Air Pollution and Daily Mortality in Detroit"
"Air Pollution and Mortality in Barcelona"
"Particulate Air Pollution and Daily Mortality in Steubenville, Ohio"
"Air Pollution and Daily Mortality in London: 1987–92"
"Air Pollution and Daily Mortality in Philadelphia"
"PM_{10} Exposure, Gaseous Pollutants, and Daily Mortality in Inchon, South Korea"
"Daily Mortality and 'Winter Type' Air Pollution in Athens, Greece"
"Air Pollution and Daily Mortality in Residential Areas of Beijing"

There is a remarkable, almost numbing, uniformity to the titles. They share a syntactic structure, differing from one another through a paradigmatic substitution of terms within that structure. In each, a compound subject is first offered through a conjunction of air pollution with mortality, later to be positioned through a locating "in." Though there are minor variations in the first half of the titles — "air pollution" might be modified as "particulate air pollution" or "winter type air pollution" — the most significant transformations take place in the second half, the prepositional phrase naming a particularity of place.

In this structure we discern something about the workings of exemplarity as political method. Through the mustering of a network of almost identical examples, and by giving their articles almost identical names, the doctors make Hong Kong an example of a much larger problem. At the same time as that example draws power from the network, it also lends stability to that network. The co-examples as a whole, as a network, substantiate a conjunction of objects — air pollution and death — differentiated only by place.

One thing to notice here is the play of particularity in the formation of political substance. Rather than jeopardizing its stability, the proliferation and accumulation of particulars is key to the citational network's existence. The production of Hong Kong air is both a localizing and a globalizing project. Localizing because it carves out the uniqueness of Hong Kong. It lends it specificity; the hallmark of that last prepositional phrase is place-based specificity. Globalizing because it performs membership in an international community of atmospheric and medical science and in an international, global problem.

Equally important, the common form of the titles signals common method. Both articles were "retrospective ecological studies" employing "time series analysis," a method that amounts to statistically correlating the "number of people dying on a particular day" (or a day or two later) with meteorological data and air pollutant concentrations over a long-term period.[7] The statistical method used was a Poisson regression model "constructed in accordance with the air pollution and health: the European approach (APHEA) protocol."[8] The near identity of the titles in this particular citational network, in other words, is premised upon a near identity of technique. It is not enough to assert that Hong Kong's deadly air is one example among many in the world; co-exemplarity is actualized through the standardization of technique.[9]

This simultaneous evocation of general problem and specificity is resonant with dynamics in other spheres. It bears comparing, for instance, with the collecting, formatting, and iterating of data in environmental informatics, creating a general problem precisely by arraying and juxtaposing particularities. As Kim Fortun observes, however, environmental informatics enables this process to be iterated across a range of sites and types of information that would be impossible to encompass in the space of a single study.[10] The Hong Kong daily mortality studies discussed here would be but two among a vast library of data sets for information engineers like those Fortun describes, raising the question of the extent to which such studies might be produced in anticipation of themselves being informatted and networked. The carving out of specificity through geographic location also underscores Sheila Jasanoff's observation that specificity plays a vital role today in legitimating claims of intellectual innovation and ownership.[11] What becomes clear looking across these topoi is that while specificity is at play in all these moments, one cannot take for granted what specificity means. There is no specificity in general, and the real work of specificity must be gleaned from the pragmatics of the specific knowledge practices in which specificity as a concept is figured.

To understand this, let us examine the specific conditions in which the citations appear in the Hong Kong articles. Consider this excerpt from the HKU study's conclusion:

> In setting air pollution control policy from a public health viewpoint, it is important to identify the health effects of air pollutants from local data. Because of the lack of data, there are few studies based on daily hospital admissions and mortality in the Asian Pacific region. For hospital admis-

sions there has been only one study in Australia (36) and two in Hong Kong (30,37). For mortality studies, there have been one in Beijing, China (38) based on 1-year daily data, two in Australia (36,39), and two in Korea (40,41). Our report should contribute to the understanding of the effects of air pollutants in this region and may clarify the differences in effects and mechanisms between Western and Eastern populations.

Local data on health effects of air pollution are required for setting standards and objectives for air pollution controls. *When local data are not available, foreign data may be helpful, but they may not be relevant or applicable because of a difference in climate or other conditions.* Our findings in this study provide information to support a review of air quality objectives with consideration of their effects on health.[12]

Here, the network (plotted by the integers corresponding with the citations at the end of the article) is invoked through a naming of its holes, "the lack of data." The naming of the general problem is indistinguishable from the claim for the primacy of the specific.

The explicit value of the Hong Kong study is marked as clarifying differences in effects "between Western and Eastern populations." By identifying Hong Kong's warm, humid summer and cool, dry winter, the HKU study reminds us that we are in the subtropics; and the specific ways in which it cites its network of relevant citations give that reminder a certain freight. It identifies and locates the work of Hong Kong doctors within the terms of a center and periphery of scientific practice. As scientists in the periphery, the researchers must negotiate a double bind not unlike the one Lawrence Cohen describes facing gerontological organizations and authors in India in the 1970s, who, in appending "India" to their names and publication titles "claim[ed] local autonomy from internationalist [gerontological] discourse, but [did] so through a reassertion of epistemological subordination."[13]

The Hong Kong doctors navigate this bind through an appeal to local appropriateness: "When local data are not available, foreign data may be helpful, but they may not be relevant or applicable because of a difference in climate or other conditions." Note, they do not say that the category does not apply "here" or that air pollution is a Western problem; they simply maintain that better, more local data is needed. This is a supplementary strategy, one that has the potential to disturb, even while leaning upon, the centrality of temperate studies: "This study provides additional information for our previous study on hospital admissions (21), and the many time series studies on air pollution and mortality in temperate countries (1–11,13,15,17–19,28,29,33,35,38,39)."[14]

The Hong Kong studies "contribute to" and provide "additional information for" the networked assemblage of other conjunctions of air and mortality "in temperate countries," and in doing so, they help it grow. Yet at the same time, their act of "adding to" articulates through implication an inadequacy in the apparently whole original to which they contribute.[15] The Hong Kong doctors' exemplification of Hong Kong names geographic unevenness in—even while extending the reach of—an emerging coalescence of scientific and political substance.

This emergent substance is fragile stuff. Daily mortality studies face criticisms that they establish no causal link or proof of impact in the long term. Some epidemiologists, for instance, argue that even if one can show that the number of people dying on a day with high air pollution is significantly greater than on a comparable day with lower pollution, the early deaths might be of people who had little time left to live anyway.[16] Those most vulnerable on high-pollution days are those with fragile health or in advanced stages of terminal illness, the argument goes. This is termed a "harvesting effect." Those who died were going to die soon; they were simply harvested early. Long-term cohort studies are needed to determine precisely how many, if any, person-years have been lost. Only with such data, this argument concludes, can the extent to which air pollution decreases life be understood.

Such a refusal to recognize air's daily effects by scaling time out seems absurd at first, but we should recognize it as a logical side effect of rendering illness and health into prognosis. As Sarah Lochlann Jain illuminates in her analysis of "living in prognosis," a prognosis—which assigns people a certain percent chance of being alive in the next number of years based on when others considered to be in comparable medical and demographic categories have died—puts one in the mind-wrenching position of living counterfactually, always juxtaposing one's living against aggregated odds of dying.[17] The analytic of harvesting simply takes this head-wrench to the extreme, by not finding a death today worthy of note simply because most others in the same position, whether good air day or bad, did not live that much longer.

Substantiating Hong Kong air as a dangerous substance will require crunching not only numbers. It will require grappling with how to think about a cause of death when causes are multiple and overlapping, and how, when lives and causes are complex, to say when it matters that a person dies today—and not tomorrow or next year. These efforts are crucial if air pollution's effects on health are to be grasped.

At the same time, they run the risk of narrowing our sense of what matters in human-atmospheric relations. When we ask how many more people die on particularly polluted days than would have if the air were clear, death becomes a proxy for air's effects, and death itself is rendered a problem of lost time—which in turn prompts the demand for more accuracy in counting the time in person-years lost. (How many person-years will be spent counting person-years?) But it bears remembering that air's human traces are found not only in those who die, their times of death, or total person-years lost, but in the fabric of living.

Air and Living

I collapsed when I got home, my stomach somersaulting like it had at the Tung Chung Citiplaza, where I had needed to stop at the public washroom instead of catching my connecting bus. A fever hit me that night, leaving me weak and useless. The next morning, I called Wong Wai King, my collaborator in Tai O, to cancel our appointment.

"You're sick, eh? Yeah, the honghei these days has been really bad."

I found this strange. The air hadn't seemed that bad. But I spoke to others, who nodded knowingly and recalled that the air had been particularly wet on that hot, muggy day.

• • •

The link forged by the doctors between air and health was not novel or isolated. Nor, as Wong Wai King and others helped me to see, was air's impact on health in Hong Kong limited to its particulate load. Already circulating was an existing discourse of honghei and health. Reviewing my notes back in San Francisco, I noticed this entry from August: "Ah Chiu has been sick. She got a cold or something. It's a common thing to get colds out here in the summer. Nobody thinks it's strange, because they all know that when going in and out of air conditioning, you can get really cold and then sick." My notes and memories are dotted with such commentaries. Sometimes, I was told, it was too hot. Other times it was cold, dry, or wet.

In traditional Chinese medicine (TCM) texts, honghei (Mandarin: kōngqì) denotes one of two sources of acquired *hei*. The other source is food. Hei, widely recognized in its Mandarin pronunciation, *qi*, is the fundamental life force in TCM, often translated as breath. Honghei is thus a breath in two senses; it is a source of vital breath, and it is breathed. In everyday use, honghei refers to the air in one's surroundings.

Though breath is vital, wind is dangerous. "Wind is the first evil," my acupuncturist back in California, Marliese, explained to me. "It opens the body to secondary ills." Historian of science Shigehisa Kuriyama offers a beautiful account of the central role played by wind (Mandarin: *feng*; Cantonese: *fung*) in the history of Chinese medical conceptions of the body. He highlights the tension that existed between, on the one hand, feelings of an ultimate resonance between the body's breath and the surrounding winds and, on the other, anxieties about human subjection to chaos, where humans were opened to irregular and volatile winds by their skin and pores. Through close study of medical and philosophical texts, Kuriyama shows clearly that "meditations on human life were once inseparable from meditations on wind," in both Chinese and Greek medicine.[18]

What most strikes me in Kuriyama's account is his attention to language—both in the ancient texts he studies and in his own writing. Wind and air whistle through his writing as much as they do through the texts he analyzes. Listen, for instance, to his discussion of the connection that the philosopher Zhuangzi drew between earthly winds and human breath.

> The winds of moral suasion, the airs that rectify the heart, and now the heavenly music of gaiety and sadness. All these bespeak a fluid, ethereal existence in a fluid, ethereal world. A living being is but a temporary concentration of breath (*qi*), death merely the scattering of this breath. There is an I, Zhuangzi assures us, a self. But this self is neither a shining Orphic soul imprisoned in the darkness of matter, nor an immaterial mind set against a material body. Anchored in neither reason nor will, it is self without essence, the site of moods and impulses whose origins are beyond reckoning, a self in which thoughts and feelings arise spontaneously, of themselves, like the winds whistling through the earth's hollows.[19]

By allowing the air to permeate his own figurations and similes, Kuriyama conveys to his readers Zhaungzi's theorization of human permeability and impermanence more vividly and viscerally than a less writerly account could.

Later Kuriyama will show how much more dangerously the winds are figured in subsequent texts, and it is this sense of wind's danger that my acupuncturist in California inherits through her study of TCM.

Air's meanings in Hong Kong seem to exceed this classical medical genealogy. Among people I have known in rural and urban Hong Kong, good fung characterizes good places. Wind's ubiquity, however, and the way it wends its way into everyday talk recall the inseparability of wind and life that Kuriyama describes and the lyrical trace of an imminently atmospheric sense of the self and health. Meditations on life through wind are as prescient as ever.[20]

• • •

In Tai O, the air is on the tip of people's tongues. "Hello, good day. Nice fung today, isn't it?" The old men sit on the benches by the Lung Tin Housing Estate, Dragon Field, facing the road that connects Tai O to the rest of Lantau Island, watching the hourly bus come in with visitors. Their shirts are loose. The breeze curls through Lung Tin, finds Wong Wai King sitting on the concrete steps outside her bottom-floor apartment. She sips some sweet water, closes her eyes, and plays her *guzheng*. "*Wah, hou shufuhk*," she tells me. "Ah, it's so very *shufuhk*."

• • •

The word *shufuhk* means "comfortable," but also more. When people say they're not shufuhk, they mean they're not well. Conversely, when Wong Wai King and others tell me that they're shufuhk, they tell me that they are experiencing a saturating pleasure. Like a cool breeze on a hot sticky day. Clean sheets on a bed. Or the way a cup of tea might warm you from the inside when you're cold. The word is ubiquitous.

Places are made into living things through a blend of landmark and language, as anthropologists of place have taught us to see, and the air in Hong Kong is undeniably part of the rhetoric of its place.[21] But air, polluted and otherwise, is a daily materiality as well as a symbolic field. To explore a material poetics of place, and air's function with it, we need to ask after the material and meaningful ways in which air enters into human and geographic life as such. For the notion of a poetics of place to have any teeth, for it to do more than simply legitimate linguistic study as a study of something linked to the material world, we must also go after the nonverbal ways air operates poetically. How does air serve as a meaningful and material unit in the building of Hong Kong? Let us take an atmosphero-poetic tour.

Some of the neighborhoods I choose for this tour are among Hong Kong's most famous. Central is the financial heart of Hong Kong and its government, whose illuminated towers, set against a foreground of the green waters of Victoria Harbour, adorn most of the stereotypical tourist images of Hong Kong. Less celebrated internationally but well known both in Hong Kong and in tourist literature is Mong Kok, a district on the Kowloon Peninsula. For many, Mong Kok is the antithesis of Central. Mong Kok is commonly held to be more Chinese than Central. While English appears on shop signs and restaurant menus in Central and sometimes comes out of shopkeepers' mouths, it is rare in Mong Kok. Whereas Central offers at least some Western comforts, Mong Kok caters to Hong Kong Chinese and to tourists seeking a flavor of Chinese alterity within Hong Kong.[22]

Tai O should be considered a part of this tour, along with Lung Kwu Tan and Ha Pak Nai. Tai O, as we saw in chapter 2, is a popular destination for domestic and foreign tourists, though not long ago it was considered a dirty backwater. Lung Kwu Tan and Ha Pak Nai, which we encountered in chapter 4, are relatively less well known villages in Hong Kong's New Territories, hemmed in by a power station and a landfill and facing the impending construction of a municipal waste incinerator. With their inclusion, another axis of difference becomes clear. Central and Mong Kok might in isolation evoke an imagined opposition between Western and Chinese in Hong Kong, but when Tai O, Lung Kwu Tan, and Ha Pak Nai become stops on our tour, Central and Mong Kok find themselves partners in urbanity set against the rural New Territories.

• • •

Central. In the winter the air in Central sweeps in dark swirls through Connaught Road, blowing under squealing double-decker trolley cars before whirling up Pedder Street toward Lan Kwai Fong, Central's famed restaurant and bar area. It chases the heels of trundling buses and racing taxis, and flings gusts of soot at the ankles of the pedestrians waiting at the crosswalk, who, almost in unison, lower their heads and cover their mouths and noses with a hand or handkerchief—a loosely synchronized nod and an almost instinctive gulp of held breath—as the wake of air washes over them.

Lung Kwu Tan. In Lung Kwu Tan and Ha Pak Nai, two villages in Hong Kong's Northwest New Territories, the air smells cleaner at first; it doesn't smell of diesel. There are fewer buses out here. Fewer taxis. But it does smell of garbage, of the garbage water that leaks from refuse trucks. People talk about the dust that settles on their vegetables from the cement factory's smokestack. Then there are the flies that fill the air, making you want to keep your mouth more tightly closed while breathing. Now residents are worried about what else might come from the air if the government builds its incinerator here. Dioxins, says Rupert, the most poisonous substance humans have ever created.

Still, the air is on the water, and this yields cool breezes. On weekends it fills the sails of windsurfers and carries the scent of visitors' barbecues, even if the occasional atmospheric shift wafts reminders of the cement factory, power station, and landfill nearby.

Mong Kok. In Mong Kok, a neighborhood on the Kowloon Peninsula that has been called the most densely populated area in the world, the winter winds are as sooty as those in Central. Dust expelled from the backs of abundant buses, trucks, and taxis barely settles before it is stirred up again. Pedestrians cross the street with the same nodding gestures as in Central. Off the street, though, the winter wind might find itself broken by a crowd, trapped and thawed by the press of people gathered to shop and play.

The same is true a bit farther north, in Yau Ma Tei, where there is also the night opera. Two women are performing, one middle-aged with glasses, leaning deliberately toward her microphone under bright incandescent lights. The musicians sit to the left, one smoking a cigarette while he plays his erhu. The music, the voice, they quaver. They sound like old radio. The air is full too, with the sticky smell of cow parts being stewed, durian, skewers of pork, oyster omelets, clams, and black beans. The scent of diesel fades into memory, and the cold air, defeated, rises to the overlooking skyscrapers in warm ripples.

· · ·

We have taken a slight detour from the issues of health that first brought us to consider the air. But we have retained the issues of the body, the question of immediacy—the coughs, the instinctive intakes of breath. Part of air's substantiability in Hong Kong comes from the fact that it is always breathed.

The poetic mattering of Hong Kong's atmosphere encompasses not only

Wong Wai King's rhapsodic "Wah, hou shufuhk," but also her sip of sweet water, the placement of her chair, and the coughs and nods of the pedestrians aiming to cross the street in Central. Air's poesis, the coproductive engagements between people and air, range from commentary, to breath, to avoidance, to the flip of an air-conditioner switch. Put another way, air is not only an object of cultural commentary, and not only a nonhuman materiality always already enmeshed in webs of social and cultural practice. It is something embodied that engages with humans through bodily practices. The smell, breath, wind, weather, typhoon, air conditioning, air pollution, height, verticality, science, sound, oxygen, smoking. The tactility of the atmosphere.

Anthropologist and musician Steven Feld has argued that sound and voice provide a useful point of entry for apprehending relations between person and place.[23] He identifies the sonic resonance of the human chest cavity as a central feature of the links and feedback loops between people and their environments. How similarly fruitful might an anthropology of air be, an anthropology of this stuff sensed in and through the moment of bringing breath into the body, or at the moment when wind opens the body to ailments? Air muddies the distinction between subjects and environments, and between subjects. This thickness and porosity rendered by air is part of what makes the air and the airborne such deeply felt elements. Bodies may be, as the geographer David Harvey argues, intersections of large- and small-scale spatial practices;[24] but if bodies are an intimate location of effects and agencies, air is the substance that bathes and ties the scales of body, region, and globe together, and that subsequently enables personal and political claims to be scaled up, to global environmental politics, and down, to the politics of health.

Air's Comparisons

In August 2000 a feature entitled "A Breath of Fresh Poison" was published in the *South China Morning Post*. In the article, readers are introduced to a sympathetic character, Fred Chan Man-hin, who had recently returned to Hong Kong from Canada to start a company. He initially "planned on being here forever," he tells the *Post*, but "the pollution has affected my decision. I can't work and be sick all the time." Today Chan "avoids his office in the Central business district because the pollution gives him dizzy spells and migraine headaches. He has spent tens of thousands of dollars on doctors and tests to find a cure for the allergies, viruses, and exhaustion that he cannot seem to shake."[25]

The article throws into relief a signature feature of air's substantiation as a problem in Hong Kong. It does not merely recount Chan's unshakable health woes; it makes a pointed comparison. Chan initially left Hong Kong for Canada, we are told, and he returned to make his fortune, but now the pollution might affect his decision to "be here forever." If air constitutes a danger in Hong Kong, part of its threat derives from its capacity to serve as an index for comparing Hong Kong with Canada and other places.

This capacity of air for comparison first became evident to me through my family, particularly through jokes about how predictably those who do not live in Hong Kong get sick when they visit. My mother's cousin, Ling, playfully chides her when she falls ill, for instance, when my parents visited Hong Kong near the end of my fieldwork. "You, your cousin Maggie, and your brother To—you all get sick whenever you come back to Hong Kong." My mother falls ill almost every time she visits Hong Kong, as do I. Ling knows this well, as we usually go to her or her husband for antibiotics. "You're not *jaahppgwaan*, not accustomed, to the air," Ling says. "Will you still visit?"

Will we still visit? This simple question draws us back to the landscape photographer's cocktail party, to my conversation with the American Chamber of Commerce representative and his wife, who wondered aloud how investors could be expected to come to Hong Kong if the air quality continued to deteriorate. It echoes Disney's admonishment to Tung. Air is not only an index of health. It is an index for comparing livability, well-being, global attractiveness.[26]

I cannot leave the matter of air's comparability at this level of global comparison, for it misses some of the subtle comparisons and distinctions that operate within the city-state. We are now acquainted with the air of some of Hong Kong's neighborhoods, its qualities, and its dangers; now questions of justice and equity beg to be asked. How are Hong Kong's air spaces distributed? Who gets to occupy those with the cleanest air? Who breathes the street? Who breathes mountains? Who breathes the sea? Who breathes flies?

• • •

A few weeks after moving to Mui Wo, I returned to Sai Ying Pun to visit with the fruit vendor, Mrs. Chau. Ah, you've come back, Mrs. Chau said, loudly enough for passersby to hear. I smiled, a bit embarrassed, and replied that the oranges looked good. I asked her to pick some for me, and for a glass of juice, and we chatted for a while there on Mui Fong Street.

I missed Sai Ying Pun, I told her. Mui Wo was nice, but it wasn't as convenient. There were also all the mosquitoes, I continued. Expecting some sympathy, I offered my arms to show her my mosquito bites, but Mrs. Chau dismissed them with a wave and a laugh.

Sure, there are mosquitoes, she said. But I'm sure the honghei is much better there.

Of course. Of course honghei mattered to Mrs. Chau, who worked every day on the busy corner of Mui Fong Street and Des Voeux Road, just down the street from one busy bus stop, where diesel buses pulled in nearly every minute, and across the street from another. Hillary, the stationer down the street, at least had a door between the street and his shop, and his shop was air-conditioned.

· · ·

Far from uniform, Hong Kong consists of pockets. Studies in the loosely Marxist or critical geographic tradition take this as an assumption—that there are social inequities, mapped and realized through spatial distinction. Through their lenses, we discern a geographically uneven distribution of environmental harm, where the rich have access to good air, while the poor are relegated to the dregs, to the smog and dust under flyovers or on the streets.[27] One can, in other words, discern a political-economic geography of air. The poorest air quality was initially in the urban areas, in the industrial zones. Now the bad air is being exported, as Hong Kong companies relocate their factories in Guangdong province on the mainland, where labor costs are lower and environmental standards more lax. But then the pollution comes back in those notorious winter winds.

These arguments help to ground the air in a solid sociological critique of social and geographic stratification; for this reason they are politically vital.[28] At the same time, such fixings need less rigid company. When mapping the spatial distribution of social inequity, an account of air must at some point leave land-based maps, for they can divert us from the movements of air and breathers alike—not to mention mobile pollution sources, such as the taxis, buses, airplanes, and cargo ships crucial to the circulations of Hong Kong's industries. To the geography of air and the dialectics of air and capital, I add three corollaries: (1) air is made not only in emissions but also in the respiration and movements of breathers; (2) neither those who emit particulate, the winds that carry it, nor those who breathe it sit still in places; and (3) as Kuriyama reminds us, there has always been more to air than particles.

. . .

The stratification of air spaces in Hong Kong has been loosely tied to income, and incomes and occupations have also been racially marked. White-collar expatriates, with their generous compensation packages, have to a greater extent than most people in Hong Kong been able to choose to live somewhere clean and central. Air spaces have been constituted in part by the racialized and classed bodies that live, work, and play in them.

The Peak and the Mid-Levels have long served Hong Kong's elite as airy refuges. Almost from the moment British colonists occupied the small island off China's southern coast, they turned toward the peaks that formed the dramatic backdrop for the harbor they so desired, looking upward for some respite from the summer heat and humidity. If for mountaineers the staggering heights of snowcapped peaks presented a dream of sublimity and transformation, the Peak in Hong Kong offered to colonists a more mundane yet perhaps equally treasured transcendence of place, time, and air.[29] Even relatively recently, civil servants have had privileged access to apartment buildings high up.

In colonial times, people cared mostly about heat and humidity. The winter winds, whose passage through the landmass of greater Asia lent them coolness and dryness, were greeted with great pleasure. Today, that dryness and that passage through China have made winter less popular than it used to be. Real estate up high continues to be prized; now, though, it is valued not only as an escape from the hot, muggy summers but also because it promises at least some relief from roadside pollution and congestion, as well as convenient access to work and play.

The Mid-Levels, known in Cantonese as *zhong saan kui*, or the "mid-mountain area," are found a bit downhill from the Peak, and they too serve as something of a refuge from the soot below. The apartment towers are spaced farther apart than in the neighborhoods at lower altitudes, and there are fewer cars. Commercial skyscrapers are less prevalent up here, and the common mode of commuting here is the longest covered outdoor escalator in the world—the same one that stars in Wong Kar-wai's film *Chungking Ex-*

press. The escalator descends into Central from the top of the Mid-Levels in the morning, carrying not only local and expatriate professionals on their way to the office, but also domestic workers heading down to the markets to buy the day's groceries. Later, at 10 a.m., the escalator will reverse itself so they won't have to climb the many flights of stairs back to their employers' homes. Scores of restaurants and bars have sprung up around the escalator. The escalator and the easy commute it offers into Central have made the terraced streets of the Mid-Levels a pocket of real estate that is even more highly valued today than it was in colonial times.

Much of Hong Kong seems designed to get off the ground—into the air, and out of it. In colonial times, the English built their mansions in the Mid-Levels and Peak. Today, when I walk with Hemen, a representative of the Tsing Tao Beer Company, he wends his way expertly through Wanchai, a government and nightlife district on Hong Kong Island, without ever touching the ground. We spend the day on the walkways that link this hotel to that shopping center. Some walkways are covered, others enclosed. Up here, we avoid the cars and the exhaust. My grandmother and I got lost once in these walkways. I remember how she pointed down to the street. There, she said, that's where I want to go. How do we get there? We never made it—we were lost in the flyovers.

. . .

Air is like food, essential to human life. Any anthropology worth its salt, however, asks after the meanings of the essential and its manifestation in material and semiotic constellations of power. Writing of food and eating, Judith Farquhar observes that "a political economy of eating emphasizes the uneven distribution of nutritional resources, while a political phenomenology of eating attends to the social practices that make an experience of eating."[30] For an adequate account, both ends of the analytic pole are necessary, as is everything in between. Air similarly calls for an understanding of its distribution and an emic analysis of its presence and distinction in acts of living. Like foods and tastes, air is enrolled in projects of social, racial, ethnic, and cultural distinction. When diasporic Chinese find the air in Hong Kong or China unbearable, their coughs, comments, and airplane tickets distinguish person and region. Consider also how atmospheric qualities figured in colonial poetics of difference.[31] The Chinese "do not suffer from the oppressive heat of the lower levels during the summer months as Europeans do," theorized the signatories to a petition in 1904 to create a "Hill District" for Europeans.[32] Air marked the moments when colonists grasped for something to concretize their deep unease—a sense that all around them, permeating everything, was difference.

Air's Index

We have seen that people in Hong Kong have a number of techniques for reading the air—dirtiness, wetness, heat, breeze, height. And we have seen how threats and health are substantiated through air's breezing and breathing. In this section, I want to look at one of the state's measures. Air's substantiations, as we have seen them thus far, present a mess for a planner or politician. To facilitate communication and policy, they need something easier to evaluate—a measure that can be translated back into coughs and particles, if need be, but that is simpler and more encapsulating. Little wonder that air, an index of so much, should have an index of its own.

The Air Pollution Index (API) in Hong Kong is calculated in a manner similar to that of other countries such as the United States, Australia, and Mexico. Air pollution monitoring stations throughout Hong Kong collect data on several target pollutants: sulfur dioxide (SO_2), carbon monoxide (CO), nitrogen dioxide (NO_2), and respirable suspended particulate (RSP). The raw data for each pollutant, usually measured in micrograms (μg) per cubic meter within a given period of time (one hour, eight hours, twenty-four hours), is turned into a subindex calibrated so that an index of 100 will correspond with a density of pollutant that is dangerous to health. That reading of 100 corresponds to different densities for different pollutants. For instance, for SO_2, an index of 100 is calibrated to 800 micrograms per cubic meter of air (800 μg/m3) in a one-hour period, while for NO_2, the 100 is calibrated to 300 μg/m3. For the general Hong Kong API, the highest of the five subindices (measured in different locations) for a given hour or day is taken as the API for that hour or day.

The clarity of the number 100—so metric!—in the index is what grabbed my attention; it brought to mind the history of the kilogram.[33] In 1799, in an effort to standardize measurements in France, the French National Assembly decreed that a "kilogram" would be defined as the mass of a decimeter of water at four degrees Celsius. Brass and platinum weights were made with equivalent mass, and the platinum one, called the Kilogramme des Archives, would eventually become the standard mass for twenty other countries in Europe through a treaty known as the Convention du Mètre. A more durable copy of the Kilogramme des Archives, made of platinum and iridium, was later fashioned as the international standard and called "K." Twenty copies of K were then apportioned to each of the signatories of the Convention du Mètre. Was this 100 of the API a universal measure, like the kilogram, calibrated across national and cultural difference through an ultimate standard?

It seems so at first. Common methods and machines internationally unite those who seek to measure air's load. These methods and machines serve as paths of translation; along them air can be turned into vials of dust, which can in turn be transformed into indices. These are "circulating references" — organizations and transformations of matter that allow material to assume more mobile forms.[34] The reversibility of these translations ensures the indices' stability and rigor, assuring their users and proponents of a pathway back to the dust. It takes an apparatus of techniques and methods — not simply the calibration of danger to the integer 100, but also the replicability and reversibility of the translations between air and number — to qualify Hong Kong's API as an index among others. There is a standardization, then, to the techniques for measurement, as well as to the form of the API.

When I reviewed the air pollution indices of several other countries, however, I was surprised to find that an API of 100 is calibrated to different amounts of dust in different places. For instance, for carbon monoxide the one-hour objective in Hong Kong is 30,000 μg/m³, while in California the equivalent objective is 23,000.[35] If the air in California had 24,000 μg/m³ of CO in it in a one-hour period, the API would read over 100 and be considered unhealthy, while in Hong Kong the API might hover only around 80 and be considered acceptable.[36] Between the final API form and the standard methods for measurements lies a space for governing what will register as risk or danger.

Most striking is the difference in objectives for RSP (PM$_{10}$). The twenty-four-hour target in Hong Kong is 180 μg/m³, while the federal standard in the United States is 150. The California standard is lower still, at 50 μg/m³, which is the same as levels deemed acceptable by the World Health Organization (WHO).[37] The Hong Kong threshold at which the air is considered to contain an unhealthy level of RSP is almost four times greater than the threshold in California. The standards for danger are different in different places.

Calibrating the API is a technique for managing the public perception of risk—for a public that includes vendors like Mrs. Chau, sick entrepreneurs like Fred Chan, corporations like Disney, and residents weighing arguments that a more democratic government could care better for its people.[38] The API can be read alongside the adjustment of risk thresholds that Joseph Dumit analyzes in the context of pharmaceutical marketing, where marketers aim to lower the published thresholds so that more people will feel unwell and, therefore, fit for medication.[39] It also has resonances with the novel iterations of data in environmental informatics explicated by Fortun.[40] Together these examples illuminate a common situation in which the ongoing tuning, tweaking, and reiterating of numbers, graphs, and maps becomes central to affective and aesthetic work—the making visible and experienceable (or invisible and unexperienceable) of risks that are difficult to articulate.[41] A symptomless biomarker becomes felt as disease, an intuited tie between social difference and health verges on presence. Through the API's calibration, the smell of diesel drifts in then out, a breath feels alternately thick and thin, clean and dirty, invigorating and debilitating. It is not simply that the API is deployed for persuasive ends, but that the technical practice of its generation—as much as commentaries on the breeze, held breaths, and treatises on the effect of southerly versus northerly winds—brings air into sense and sensibility. This is an aesthetic technology with serious stakes.

Air's Poetics

> First of all the enveloping hot air, ungiving, with not a flicker of movement,
> a still thermal from which there is no relief. You are surrounded by hot air,
> buoyed up by hot air, weighed down by hot air. You inhale hot air, you swallow
> hot air, you feel hot air behind the ears, between the legs, between the toes,
> under the feet.
> Many hours later, a very slight stir, followed by the suggestion of a breeze.
> The thermal remains.
> Yet more hours later, a sudden tearing gust of wind, and the storm has
> arrived.
> —Louise Ho, "Storm"

What kind of substance is Hong Kong's air? One shared, particular, and comparable, one realized in bodily, sensory, practical engagements of breath and movement, as well as through the material and mathematical transformations of medical method. One fixed in the whorls between buildings, mobile as it blows across town, across borders, across disciplines—one that signals a global political economy, postcolonial anxiety, as well as concerns about health and well-being.

Air's qualities are coupled with Hong Kong's industries. Think of the smokestacks of industrial factories making goods and the cargo ships moving freight; the carbon footprints of the jets and taxis moving finance workers; the mark on the air from the coal- and gas-burning power plants that send electricity to Hong Kong's skyline and to the electronics shops, bursting with gleaming toys to be bought and powered with leisure money or credit. Think of the combustion at the end of consumption's life cycle, where discarded things are incinerated. Air pollution is both condition and effect of capital. We burn in making, we burn in consuming, we burn in discarding, and the smoke has nowhere to go but up. Once up, this smoke constitutes its own threat to Hong Kong's place in financial circuits.

Hong Kong doctors, meanwhile, work to locate their concerns about the atmospheric load in Hong Kong within broader concerns about health, as well as within international science. Pedestrians and environmentalists worry about the winter shift in the wind that brings China's air into Hong Kong. Air's capacity to hold many forms of substance helped solidify a village–NGO collaboration mobilized to halt construction of an incinerator in Hong Kong's New Territories.

Air disrespects borders, yet at the same time is constituted through difference. Neighborhoods have different atmospheres; nations generate and apply different pollution standards; leaders worry about the state of their air compared to that of others. The winds themselves derive from differences in air pressure between regions, and similar relativities allow our lungs to inhale and exhale. Gradients, whose foundations are the contact and bleeding of difference, move air through the spaces we live in and through our bodies.

• • •

How do we theorize this shifting substance bound up in processes of production and consumption that also holds and touches much more? What manner of thinking about scales, distinctions, and connections does it open to us? My answers to these questions remain preliminary, but let me outline for now an argument for air's potential to reorient discussions of political universalism.

Recent efforts in post-Marxist political philosophy to retheorize universalism can be brought fruitfully to bear in the analysis of air, but they also meet a limit. As exemplars of such efforts, consider the interventions made by Butler, Laclau, and Žižek in *Contingency, Hegemony, Universality*.[42] The authors in this exchange agree that there are no obvious political or ethical universals unstained by particularity, and that the concepts of the universal and the particular are best understood in relation with each other and with their deployment in historically specific political acts. On the question of how precisely to understand the relation of the universal and the particular, however, the authors differ strongly.

For Laclau, the universal is an "impossible and necessary object" in the constitution of any political articulation, in both theoretical and political terms. "From a theoretical point of view," he argues, "the very notion of particularity presupposes that of totality . . . And, politically speaking, the right of particular groups of agents—ethnic, national or sexual minorities, for instance—can be formulated only as *universal* rights."[43] The particular is thus for Laclau never outside of, or prior to, a field of relative and necessary universality within which particulars come to be known as such. The universal, in its very impossibility and necessity, grounds the politics (and analytics) of particularity.

Butler, meanwhile, argues almost the reverse point. "If the 'particular' is actually studied in its particularity," she writes, "it may be that a certain competing version of universality is intrinsic to the particular movement itself."[44] That is, a close study of particular political movements might reveal that they actually refigure the universals that they seemed to rely upon. Universality, for Butler, rather than simply preceding the particular, is in fact generated and iterated through particular visions of the universal.

Žižek, following Hegel and Marx, invokes the concepts of oppositional determination and the concrete universal to solve the paradox of the universal and particular's simultaneity. Of all species within a genus, he argues, there is always one that is both member of the genus and determiner of the terms defining that genus. Furthermore, the historically specific condition of global capital structures the situation of political particularisms; and class politics, he maintains, while one among multiple forms of politics, serves as the model for politics in general.

Any of these positions could ground air's analysis to good effect. We might lean upon Butler's concept of "competing universalities" to argue that the daily mortalities substantiated by Hong Kong's doctors not only buttress a universalizing claim of air pollution's link with dying, but also instantiate a particular, competing version of this universality that questions the peripheralization of Hong Kong scientists and Hong Kong health in international science. We could borrow a page from Žižek to argue that in air's entanglement with capital we encounter the air relation determining all other air relations. Or, twisting somewhat Laclau's characterization of the relation between universalism and contingently articulated political blocs, we could see air emerging as an empty yet always necessary universal—to be filled in with honghei, RSP, typhoons, buses, breezes, science, flies—making environmental politics, rather than class politics, a primary field for political claims.

Before long, however, air would push back. Each approach offers a theory of politics through a solution to the universal/particular paradox; but to do so each leans upon an initial opposition between the universal and the particular to render their coexistence paradoxical in the first place, in need of a solution. As I hope to have conveyed, however, air's encompassment of universal and particular does not present itself as a paradox. It is a banality. Rather than a solution to a paradox of scale, then, air asks for a theoretical language that does not find its movement through multiple scales and political forms remarkable in the first place.

Can we, following Kuriyama, learn to hear air whistling through the hollows of theory? Doing so means making permeable the grounding distinction drawn between the unruly manifold of matter and putatively prior conceptual forms.[45] For ethnography, it also means adopting a different relationship than usual with the concrete. Listening to air, thinking through this diffuse stuff in the thick of becoming, requires less literal materialism.

This reminds me of the remarks of Charles Bernstein, a poet and theorist of poetics, on the relation between poetry and philosophy: "Poetry is the trump; that is to say, in my philosophy, poetry has the power to absorb these other forms of writing, but these other forms do not have that power over poetry. . . . When I think of the relation of poetry to philosophy, I'm always thinking of the poeticizing of philosophy, or making the poetic thinking that is involved in philosophy more explicit."[46] Thinking, for Bernstein, is always a poetic act. Poetry is always thinking. This figuring of always poeticized philosophy pushes me to make explicit the poetic thinking involved in theorizing problems of universality and scale.[47] What are the "universal" and "particular" but conventionalized figures for theory's poetics? Their ossification should be clear when those most ardently debating their definition declare the inadequacy of their terms, and then return to rest on them again and again. Some tropic invigoration might help—a poetic revival through the activation of examples, where details yield not simply particularity but the potential for mobile metaphors. Might the material poetics of the substantiations of Hong Kong's air—with its whirlings, its blowing through scales and borders, its condensations, its physical engagements, its freight of colonial, economic, and bodily worries about health and well-being, its capacity to link and to divide, its harnessing for simultaneously local and cosmopolitan projects—provide that reviving breath theory needs?

Chapter 1. Problems of a Political Nature

1 A pseudonym, like most names in this book.

2 Boas, "The Limitations of the Comparative Method of Anthropology." For a playful but provocative rereading of the promise of comparative method through matsutake mushrooms and the work of Marilyn Strathern, see Tsing, "Kinship and Science in the Genus Tricholoma."

3 See Anderson, *The Spectre of Comparisons*; Cheah and Culler, *Grounds of Comparison*.

4 The FOE Hong Kong office later became estranged from the international FOE network for accepting money from a major corporation.

5 On the advent of planetary imagination in and through environmental politics, see Ingold, "Globes and Spheres"; Jasanoff, "Heaven and Earth." For important attention to the colonial life of planetary consciousness, see Pratt, "Science, Planetary Consciousness, Interiors"; Grove, *Green Imperialism*.

6 Lai, "Greening of Hong Kong?" 268. For a resonant consideration of how the Chernobyl disaster sparked in the Ukraine critiques of Soviet governance in simultaneously environmentalist and nationalist idioms, see Petryna, *Life Exposed*.

7 Exemplary works on the commingled constitution of nature and culture include Haraway, *Primate Visions*; Latour, *We Have Never Been Modern*; Raffles, *In Amazonia*.

8 For an excellent account of the various meanings that the concept of the "ecosystem" has denoted and some of the struggles that have transpired within the field of ecology to define the term, see Golley, *A History of the Ecosystem Concept in Ecology*.

9 Problems of comparison take center stage in postcolonial and transnational science studies. See Langwick, *Bodies, Politics and African Healing*; Lock, *Twice Dead*; Lowe, *Wild Profusion*; Zhan, "Does It Takes a Miracle?"

10 For an incisive critique of the analytic of "transition" in China Studies, see Zhan, "Civet Cats, Fried Grasshoppers, and David Beckham's Pajamas."

11 My thanks to an anonymous reader for helping me to clarify this point.

12 Kim Fortun, in *Advocacy after Bhopal*, writes beautifully of her own need to act while researching the aftermath of the Union Carbide chemical spill in Bhopal, India. Her analysis of the pragmatics of writing advocacy materials employing genres of expertise that one might at other moments criticize illuminates the multiple, often contradictory, political planes that activists and knowledge producers negotiate. See also Sawyer's *Crude Chronicles*, a compelling study of the transnational politics of indigeneity and class catalyzed by petroleum capital in Ecuador. Sawyer adopts a parallel, though slightly different, analytic stance, grounding herself and her research in the urgency of political struggle.

Chapter 2. Endangerment

1 That some fauna have more appeal than others in conservation circles is well known but rarely theorized. For a notable exception, see Lorimer, "Nonhuman Charisma."

2 Perhaps anticipating a negative response to such a recommendation, the planners made their proposal in roundabout fashion: "In recognition of the stilted structures being an essential component of Tai O's heritage and fishing-village character, a mixed approach of retaining the stilted structures with removal of the minimum necessary for the construction of river walls along Tai O Creek is proposed. The stilted housing area in the downstream area of the creek will be removed and could be rebuilt in a new stilted form of architecture for commercial and visitor uses." Notice how in the proposal the removal of homes happens *passively*, without any agent of destruction. The proposal is first buffered by a clause that recognizes the importance of "stilted structures" to Tai O's heritage and character, then hidden in the euphemism of a "mixed approach," in which not demolishing *every single* stilt home counts as a mixture of "retaining" and "removal." Furthermore, a misleading suggestion of *renewal* rather than *destruction* inheres in the word "rebuilt," which hastily follows the proposal to remove stilted housing along Tai O Creek. I say the suggestion is misleading because what is rebuilt is not "stilted housing," but "a new stilted form of architecture for commercial and visitor use." The function of the stilted language becomes clear: within its logic, even if long-standing homes are removed, Tai O's character can be considered to be safeguarded so long as there exists some form of building on stilts. The discourse recognizes "stilted structures"—but not stilted housing—as essential components of Tai O's heritage and fishing-village character.

3 Liu and Hills, "Environmental Planning, Biodiversity and the Development Process," 358.

4 Clarke, Jackson, and Neff, "Development of a Risk Assessment Methodology for

Evaluating Potential Impacts Associated with Contaminated Mud Disposal in the Marine Environment," 69.

5 Man, "The Environment," 346.

6 Ibid., 346, emphasis added.

7 Guldin, "Hong Kong Ethnicity."

8 Literally meaning "ghost person," *gwailou* has a history of racializing use dating back to early encounters between Chinese and European travelers. The term ranges widely in function today, depending on context, speaker, and audience—from angry derogation to teasing affection and self-deprecatory identification.

9 Mathews, "Hèunggóngyàhn."

10 Interestingly, in the history of the Chinese language, the very notion of a Chinese character and Chinese culture (*wenhua*) was itself developed as a means to assert distinction. It was never a contextless and neutral self-description. Lydia Liu, in *Translingual Practice*, links this development to the history of "East-West encounter."

11 Wong, *Tai O: Love Stories of the Fishing Village.*

12 On the discursive, historic, epistemic, technological, cultural, habitual, and other enmeshments through which our bodies and senses (and even our senses of our bodies) emerge in the first place, see Butler, *Bodies That Matter*; Dumit, *Picturing Personhood*; Foucault, *Discipline and Punish*; Langwick, *Bodies, Politics and African Healing*; Massumi, *Parables for the Virtual*. Also useful for me in thinking about the relation between sensory and environmental politics has been work on the poetics of place. *Senses of Place* (Feld and Basso) is an exemplar of this tradition, analytically weaving physical spaces with vernacular textures with remarkable grace.

13 Ivy, *Discourses of the Vanishing.*

14 Rosaldo, *Culture and Truth.*

15 Ma, "Re-advertising Hong Kong," 140, 147.

16 Ibid., 156.

17 Stewart, "Nostalgia—A Polemic," 227.

18 Rofel, in *Other Modernities*, considers the nostalgia that older women workers in China have for times past when their labor merited socialist and nationalist pride. She writes, "Nostalgia is palpable in the way the stories jar against everything happening in the world around them: changing gender differentiations that make them appear to be women who misperceive what it really means to be a woman rather than heroically overcoming the need to be merely women; the devaluation of workers that makes their labor appear to be just common drudgery rather than Herculean toil for the nation; the feminization of labor in state-run work units that makes them appear to be marginal to the new story" (144).

19 For an excellent analysis of colonial land law in Hong Kong, see Chun, *Unstructuring Chinese Society.*

Chapter 3. Specific Life

1   Hu and Barretto, "New Species and Varieties of Orchidaceae in Hong Kong," 6.
2   Ibid., 1, 4, 34, 2.
3   Ibid., 22.
4   Ibid., 4.
5   Ibid., 6.
6   Sun, "The Allopolyploid Origin of Spiranthes hongkongensis (Orchidaceae)."
7   Mathews, "Hèunggóngyàhn."
8   See Tam, "Eating Metropolitaneity" and "Heunggongyan Forever."
9   Watson, Golden Arches East.
10 Urban, Metaculture.
11 See, for instance, Ho, "Situating Global Capitalisms"; Maurer, Mutual Life, Limited; Mitchell, Rule of Experts; Miyazaki, "The Temporalities of the Market"; Riles, Collateral Knowledge.
12 Cheah, Spectral Nationality. For Lenoir's work on Blumenbach, see The Strategy of Life.
13 Contemporary anthropologists might contest the truth of this universality of human categories, but the grip that Kant's thought holds in political theory, and the almost complete universalization of it in the shape of modern political science, is incontestable.
14 Hong Kong Research Grants Council, Annual Report 1999, 41.
15 For orchids, this is a valid statement, as the intense speciation of orchids produce many small populations of different species.
16 This dominance of genomic life is conditioned by several technopolitical developments: first, the development of techniques for transcribing DNA into code; and second, directly related to the first, the rush in the biotechnology industry to specify and patent unique strings of code. For some excellent discussions, see Haraway, Modest_Witness@Second_Millennium; Hayden, When Nature Goes Public; Rabinow, French DNA; Reardon, Race to the Finish; Sunder Rajan, Biocapital.
17 This relative eclipse of ecological life by genetic life has occurred perhaps because biotechnology, as Kaushik Sunder Rajan argues, coevolved in such a striking and public fashion with the promissory claims of highly capitalized entrepreneurial and speculative enterprises. See Biocapital.
18 Abbas, Hong Kong, 6.
19 Sharon Traweek's analysis, in Beamtimes and Lifetimes, of the "culture of no culture" of American high-energy physicists, informs my discussion here.
20 Abbas, Hong Kong, 6–7, emphasis added.
21 Law, "Northbound Colonialism," 203.
22 Ibid. See also Man and Lo, Cultural Identities and Cultural Politics.

23 The newspaper article was later reposted on a website dedicated to contesting Article 23, http://www.againstarticle23.org.

24 Gillmor, "Proposed Security Law Putting Liberties at Risk in Hong Kong."

25 For a thorough account of English-language, Chinese, and Japanese media coverage of the handover, see Lee et al., *Global Media Spectacle*.

26 See Golley, *A History of the Ecosystem Concept in Ecology*. For a provocative study of the divergent values attributed to complexity, holism, organism, and system by different thinkers in the history of ecology, see Kwa, "Romantic and Baroque Conceptions of Complex Wholes in the Sciences."

27 According to Hong Kong government figures, the number of Chinese factory workers employed by Hong Kong companies has skyrocketed in the past twenty to thirty years. "In 1981, Hong Kong companies employed approximately 870,000 manufacturing workers in Hong Kong and few elsewhere. By 2002, Hong Kong companies employed fewer than 200,000 manufacturing workers in Hong Kong, but between 10 million and 11 million in the Pearl River Delta region" (InvestHK, Government of Hong Kong, "Hong Kong's Links with the Delta," http://www .investhk.gov.hk/). By 2009, twelve million Chinese workers were employed by Hong Kong companies in Guangdong province. InvestHK, Government of Hong Kong, "Hong Kong's Unrivalled Location," http://www.investhk.gov.hk/.

28 InvestHK, Government of Hong Kong, "CEPA," http://www.investhk.gov.hk/.

Chapter 4. Articulated Knowledges

1 There is an excellent body of work on the anthropology of place. See Rodman, "Empowering Place." Key works highlight the spatial dimensions of both everyday and more dramatic political struggles. See, for instance, Gupta and Ferguson, *Culture, Power, Place*; Low and Lawrence-Zúñiga, *The Anthropology of Space and Place*. On the poetics of place, see Feld and Basso, *Senses of Place*; Stewart, *A Space on the Side of the Road*. Also see Brown, *Dropping Anchor, Setting Sail*; Raffles, *In Amazonia*.

2 See Collins and Pinch, *The Golem*; Fleck, *Genesis and Development of a Scientific Fact*; Shapin and Schaffer, *Leviathan and the Air-Pump*.

3 Exemplary ethnographies of science include Helmreich, *Silicon Second Nature*; Martin, *Flexible Bodies* and *The Woman in the Body*; Rabinow, *Making PCR*; and Rapp, *Testing Women, Testing the Fetus*. Researchers have also extended the anthropology of science to include postcolonial and international political-economic power relations. See Abraham, *The Making of the Indian Atomic Bomb*; Gupta, *Postcolonial Developments*; Hayden, *When Nature Goes Public*; Lowe, *Wild Profusion*; Petryna, *Life Exposed*.

4 The concept of "local knowledge" has obvious ties to the history of anthropology in the United States (for example, Clifford Geertz's classic essay, "Local Knowledge"), but it has by now become an almost naturalized common sense category

in environmental politics. For a book-length example of the environmentalist use of the concept, see Brush and Stabinsky, *Valuing Local Knowledge*. More nuanced studies of interacting knowledges include Gupta, *Postcolonial Developments*; Langwick, "Articulate(d) Bodies" and *Bodies, Politics and African Healing*; Lowe, "Making the Monkey."

5 On the effects of technocracy, see Jasanoff, *The Fifth Branch* and "(No?) Accounting for Expertise"; Mitchell, *Rule of Experts*. Kim Fortun suggests an analysis of advocacy as a method of (counter)expertise in "The Bhopal Disaster" and *Advocacy after Bhopal*.

6 The promises and pitfalls of political universals have been debated, for instance, in feminist theory, Marxist and post-Marxist theory, and studies of nationalism and cosmopolitanism. See Mohanty, Russo, and Torres, *Third World Women and the Politics of Feminism*; Gibson-Graham, *The End of Capitalism*; Butler, Laclau, and Žižek, *Contingency, Hegemony, Universality*; Cheah, *Spectral Nationality*; Cheah and Robbins, *Cosmopolitics*.

7 This point is formulated especially well by Anna Tsing in *Friction*.

8 On the need to revise anthropology's concepts when encountering parallel concepts in one's objects of study, see Maurer, "Anthropological and Accounting Knowledge in Islamic Banking and Finance"; Miyazaki and Riles, "Failure as an Endpoint"; Riles, *The Network Inside Out*; Strathern, *Partial Connections*. Tsing makes a related argument in *Friction* about universals, emphasizing the need to address their sticky, practical engagements.

9 See Shapin and Schaffer, *Leviathan and the Air-Pump*.

10 See Haraway, *Modest_Witness@Second_Millennium*.

11 Lorraine Daston has called this an image of objectivity as "aperspectival objectivity," where objective knowledge is unnecessarily conflated with knowledge that comes from no particular person or place. See Daston, "Objectivity and the Escape from Perspective."

12 On scientific expertise as testimony, see Golan, *Laws of Men and Laws of Nature*.

13 Conversation with author, November 1999.

14 Conversation with author, November 1999.

15 Kurzman, "Embodiment and Ability."

16 Agarwal and Narain, *Global Warming in an Unequal World*.

17 My thinking in this section is indebted to conversations with Cori Hayden and Marina Welker. See also Hayden, *When Nature Goes Public*; Welker, "'Corporate Security Begins in the Community.'" Other significant works addressing the politics and analytics of accountability, transparency, and participation include Cooke and Kothari, *Participation*; Maurer, "Anthropological and Accounting Knowledge in Islamic Banking and Finance"; Power, *The Audit Society*; and Strathern, *Audit Cultures*.

18 See Jasanoff and Martello, *Earthly Politics*.

19 See Agarwal and Narain, *Global Warming in an Unequal World*. Also see Escobar, "After Nature"; Sawyer and Agrawal, "Environmental Orientalisms."

20 See Gould, Schnaiberg, and Weinberg, *Local Environmental Struggles*. On citizen expertise, see Di Chiro, "Defining Environmental Justice." On "working knowledge," see Fortun's recent work on environmental toxicology, "Biopolitics and the Informating of Environmentalism."

21 On the limits of calls for local knowledge and participation, see Covey, "Accountability and Effectiveness in NGO Policy Alliances"; Forsyth, "Social Movements and Environmental Democratization in Thailand."

22 The concept of "articulation" comes from Laclau and Mouffe, *Hegemony and Socialist Strategy*.

23 For more on the powers and limits of "stakeholder" models in environmental politics, as well as an important alternative theorization of "enunciatory communities," see Fortun, *Advocacy after Bhopal*. See also Haraway, "Situated Knowledges."

24 The argument here is meant to resonate with Tsing's analysis of "natural translations" and articulations that bring unlikely collaborators together in environmental politics; see "Transitions as Translations." For more on surprising coalescences, see Kirksey, "Foam Frogs and Eco-Tractors."

My aim is to push the idea of translation still further, to attend not only to the fact that meanings and stakes must be translated across social worlds in collaborative politics but also to analyze the translation event itself as a speech event or linguistic technology that produces articulative collaboration, expert authority, and different kinds of subjects. This last point is similar in spirit to Fortun's attention to the emergence of enunciatory communities in acts of advocacy; see *Advocacy after Bhopal*.

25 See Callon, "Some Elements of a Sociology of Translation"; Latour, *Science in Action* and *Aramis, or the Love of Technology*; Star and Griesemer, "Institutional Ecology, 'Translations' and Boundary Objects."

26 See Langwick, "Devils, Parasites, and Fierce Needles"; Pigg, "Languages of Sex and AIDS in Nepal." For a fascinating analysis of repeated language that brackets the usual worries over linguistic equivalence in favor of an account of the culturally generative effects of "dubbed" culture, see Boellstorff, "Dubbing Culture."

27 Maurer, following Silverstein, makes a similar call for attention to metapragmatics in his accounting of accounting: "Anthropological and Accounting Knowledge in Islamic Banking and Finance"; Silverstein, "Shifters, Linguistic Categories, and Cultural Description."

28 See Bauman, "Mediational Performance, Traditionalization, and the Authorization of Discourse" and *A World of Others' Words*.

29 For an interesting analysis of ritual in professional audit practices, see Harper, "The Social Organization of the IMF's Mission Work." Harper's suggestion that "it is partly through ritual that the symbolic importance of the events are demon-

strated and achieved" (24), dovetails especially well with the point I seek to make here concerning the work of this scene as a particular form of speech event.

30 Hall and Grossberg, "On Postmodernism and Articulation," 141.

31 The concept of articulation was particularly salient in efforts to revise certain assumptions of Marxist theory, particularly the view that those social collectivities structured by class are more self-evident, inevitable, or crucial than others. Laclau and Mouffe maintained that contrary to such orthodoxy, crucial action might be taken by collectivities built around other modes of identification that are neither more nor less fundamental to the social structure. Laclau, Mouffe, and others took articulation to refer to a process whereby contingent collectivity is discursively produced in such mobilizations. For further explanation, see Laclau and Mouffe, *Hegemony and Socialist Strategy*, 96–114. The concept has been revisited recently in Butler, Laclau, and Žižek, *Contingency, Hegemony, Universality*.

32 See Li, "Articulating Indigenous Identity in Indonesia"; Moore, "Contesting Terrain in Zimbabwe's Eastern Highlands" and "The Crucible of Cultural Politics"; Tsing, *Friction*, "The Global Situation," and "Transitions as Translations."

33 Moore, Kosek, and Pandian, "Introduction: The Cultural Politics of Race and Nature."

34 Tsing, "Inside the Economy of Appearances."

35 Li, "Articulating Indigenous Identity in Indonesia," 169.

36 Translation as linkage has received considerable attention in science studies, particularly in works shaped by actor-network theory. Exemplary texts include Callon, "Some Elements of a Sociology of Translation"; Latour, *Science in Action*. For a related but different approach, see Star and Griesemer, "Institutional Ecology, 'Translations' and Boundary Objects." I push translation and articulation together here in order to mark the importance of linking science studies' attention to translation and knowledge production with an agonistic account of politics as an ongoing struggle for position within a given political field or the very coordinates through which the field is conceptualized. This is an effort to develop within science studies an alternative to the liberal democratic vocabulary associated with many analyses of scientific expertise and decision making, one attentive to struggles for hegemony, as well as an effort to inform political theorization with a greater attention to the material practices of knowledge production.

As an anonymous reviewer has helped me to see, my thinking about hegemony and articulation implicitly traces a particular genealogy through poststructuralist readings of Gramsci. See Butler, Laclau, and Žižek, *Contingency, Hegemony, Universality*; and Laclau and Mouffe, *Hegemony and Socialist Strategy*. These readings depart somewhat from prior, arguably more faithful, readings of Gramsci that placed more emphasis on the role of individual leadership in the struggle for hegemony, including the work of organic intellectuals. On the role of individual leadership in Gramsci's theorization of hegemony, see Kurtz, "Hegemony and Anthropology."

I address individuals' work somewhat differently in attending to the formation of environmentalist subjects (see chapter 5).

37 Butler, "Competing Universalities."

38 On other unanticipated outcomes of cross-cultural translation, see Liu, *Translingual Practice*; Pratt, *Imperial Eyes*; Rafael, *Contracting Colonialism*.

39 On enunciatory communities, see Fortun, *Advocacy after Bhopal*.

40 Conversation with author, January 10, 2000.

41 James Scott's work on "hidden transcripts" and "ideological resistance" is exemplary; see *Weapons of the Weak* and *Domination and the Arts of Resistance*. Kathleen Stewart's study of life in a West Virginia coal town is exemplary as well, but her work explores how critique is embodied in the texture of discourse as much as in its content, and the affective forces and potentials that might be held in people's ruminations and stories; see *A Space on the Side of the Road*.

42 Charles Briggs has also made a compelling argument for adding to science studies not only an account of counterknowledge, or "counternarratives," but also an analysis of the political-economic relations that enable or limit the transformation of such narratives into public, politically available discourse; see "Theorizing Modernity Conspiratorially."

43 Conversation with author, January 10, 2000.

44 This discussion is indebted to Allen Chun's excellent analysis of colonial land administration in Hong Kong in *Unstructuring Chinese Society*. Chun goes beyond a critique of the reification of custom; he asks after the shared compulsion to understand and fix social structure in the social sciences and colonial administration. Also see Chan, "Politicizing Tradition."

45 The gendering limitations of the ordinance have been strongly contested. See Chan, "Negotiating Tradition"; Merry and Stern, "The Female Inheritance Movement in Hong Kong."

46 Kuletz, *The Tainted Desert*; Pellow, *Garbage Wars*; Szasz, *Ecopopulism*.

47 Jasanoff and Martello, *Earthly Politics*.

Chapter 5. Earthly Vocations

1 In this, the discourse on environmental awareness parallels discussions of global sex education. See Adams and Pigg, *Sex in Development*.

2 A number of studies have illuminated what can be quite disparate itineraries and investments built into particular formations of environmentalist subjects. See Agrawal, *Environmentality*; Kosek, *Understories*; Lowe, *Wild Profusion*; Sturgeon, *Ecofeminist Natures*; Tsing, *Friction*.

3 See Cheah and Robbins, *Cosmopolitics*.

4 Cheah, "Introduction Part II: The Cosmopolitical—Today."

5 See Kaplan, *Questions of Travel*.

6 Ong, "Cyberpublics and Diaspora Politics among Transnational Chinese."

7 See Clifford, "Mixed Feelings."

8 My use of the term *vocation* also points to the entanglements of ideals, commitments, and practice and to a life and self as something cultivated. A wonderful account and theorization of the politics of such cultivation can be found in Mahmood, *Politics of Piety*. Such attention to practice and cultivation as itself a vocation and politics offers a way around the usual either/or questions of whether certain acts and stances are conditioned "externally" or come from a truer internal compass. See also Weber, *The Vocation Lectures*.

9 My thinking on conviction as interpellation is indebted to Susan Harding's account of conviction in American evangelical Christianity; see "Convicted by the Holy Spirit" and *The Book of Jerry Falwell*.

10 See Adler, "Cultivating Wilderness"; Pratt, "Alexander von Humboldt and the Reinvention of América"; Taylor, "The Tributaries of Radical Environmentalism."

11 Badiou, *Saint Paul*. The context for Badiou's turn to Paul is his perception of a crisis of political faith—faith in Marxism, in particular—and the apparent demise of militant political universalism as particularized political identities increasingly saturate political imagination. Can we, he wonders, conceive a universalism that is not grounded in facts of the present day, and thus push our political imagination beyond particular political problems to conceive of, and to act to right, problems of a more systemic nature? Can we recuperate a politics of the universal?

12 For Badiou, political commitment is inseparable from universalism, and Saint Paul founded the concept of universalism as we know it. Paul did this through his constant declaration throughout his travels that all social differences and all criteria for distinguishing people and claims were obviated by the truth of the Resurrection. That truth was indifferent to their differences, and Paul's commitment to it made possible the conception of such an indifference. In a similar vein, Slavoj Žižek provocatively celebrates Lenin as Marx's most effective formalizer, one whose legacy is a politics of truth. See Brown, "$\{\varnothing,\$\} \in \{\$\}$?"; Žižek, *Repeating Lenin*.

13 Like Isabelle Stengers, in *Cosmopolitiques*, I see more potential in a less adamant psychosocial type for science and politics.

14 For an excellent analysis of the cultural and historical conditions underlying the recent rise of Filipina domestic labor in Hong Kong, see Constable, *Maid to Order in Hong Kong*. Constable also offers a perceptive account of jokes and other practices through which Filipina domestic workers negotiate what are often racially charged work and living environments.

15 For an insightful discussion of the controversies surrounding the construction of "monster houses" in Vancouver by new Hong Kong immigrants, see Mitchell, "Transnational Subjects."

16 Shiho Satsuka, in "Traveling Nature, Imagining the Globe," attends to the ways

wilderness discourse has functioned as a nationalistic discourse in Canada in her study of encounters between Japanese tourists and Canadian environmentalists in Canada's national parks. Canadian environmental discourse and tourist materials each take up Canada's dramatic mountains and forests as symbols of a national ruggedness, unmarked by the signs of corruption or materialism that characterize the United States—a ruggedness that therefore functions as proof of distinction from its less natural, more problematically political neighbor. See also Braun, *The Intemperate Rainforest*.

17 Investment incentives were offered to potential immigrants to Vancouver; this was part of a plan to build Vancouver as a Pacific Rim city. See Mitchell, "Transnational Subjects."

18 Later I learned this was standard procedure for protest actions. Police rarely made arrests but always asked for participants' Hong Kong ID. All Hong Kong residents are required to carry it on their person at all times. I regularly witnessed police stopping people on the street to ask for their IDs and was stopped myself on several occasions. I suspect the practice was not unrelated to the then widespread concern in Hong Kong about illegal immigration from mainland China and the prospect of masses of people from the People's Republic gaining "right of abode." Some of this concern was unfounded; it was discovered that the Immigration Department had inflated figures to convey a greater sense of urgency for immigration restriction measures.

19 Sturgeon, *Ecofeminist Natures*, 50, 51.

20 What do we make, if anything, of the fact that the phrase *hou man* uses the English word "man"? One could suggest that, in addition to being funny, the code-switch implies that the manly endeavor being commented upon reflects a notion of masculinity originally located outside of Hong Kong. Such an argument is tempting, but it relies on a problematic assumption that the appearance of English lexemes in Cantonese discourse necessarily connotes foreignness. Certainly, they point "beyond" Hong Kong in many cases, but insertions of English words are often used simply for poetic and comic effect and recently have even been taken as signs of uniquely Hong Kong ways of talking. In these cases they are not beyond Hong Kong at all but specific to it.

Superb examples of this kind of language-play can be found in Toe Yuen's animated film *Mak Dau Goo Si* (My life as McDull), which won the Critics' Prize at the Hong Kong International Film Festival in 2002 and garnered accolades for "being both local [to Hong Kong] and universal." The film merits extensive study on its own terms, but to get a sense of the Cantonese-English games, one need watch only the first ten minutes. While the protagonist, child-pig McDull, introduces himself through a sing-over, fluffy clouds fill the visual field and change shape to illustrate his song. When McDull first starts to sing, the clouds change shape to resemble a pig wearing a bunny suit. The fun begins when McDull sings at some point, "Ngo dak, dak, dak" (I can do it, I can do it)—the clouds take on the shape of a duck. McDull responds happily by singing in Cantonese about how much he likes to eat duck (ngaap) and chicken (gai), and the clouds oblige by switching rapidly between fowl forms, this time cued by the Cantonese words. McDull's name is itself a pun on the pan or dau that watches over the baby pig's birth while magically flying by.

21 I was also fascinated and perplexed by a tale of two engineers who lost their oars in the middle of a lake filled with toxic waste. They never told me how they made it to shore.

22 See Ong, Flexible Citizenship, especially chapters 2 and 4. Ong convincingly argues that the figure of a humble, neo-Confucian merchant pervades international business arenas and underlies the increased visibility and acceptance of ethnic Chinese capitalists in Southeast Asian countries where they are an ethnic minority. The romance of the mobile and modest Chinese merchant is propagated in forums of international business—for instance, through Forbes magazine's description of tycoons "such as Indonesia's Liem [Sioe Liong], Malaysia's [Richard] Kuok, and Hong Kong's Li Ka-shing" as having "the same image: trustworthy, loyal, humble, gentlemanly, skilled at networking and willing to leave something on the table for partners" (cited in Ong, 146). This romance of an honorable, Confucian, merchant masculinity has enabled and morally legitimated a new order of alliances not only among Chinese capitalists but also between Chinese capitalists and heads of state in Southeast Asia, all the while overlooking the complicating fact that "Confucian philosophy . . . regards [traders'] singular pursuit of wealth as the very antithesis of Confucian values" (145).

Chapter 6. Air's Substantiations

1 Xi Xi, Marvels of a Floating City and Other Stories, 106.

2 Berman, All That Is Solid Melts into Air.

3 Marx, "The Communist Manifesto," 475–76.

4 For instance, see Harvey, A Brief History of Neoliberalism and The Condition of Postmodernity; Jameson, "Postmodernism, or the Cultural Logic of Late Capitalism."

5 Karl Marx, "Speech at the Anniversary of the People's Paper," 577–78. Cited in Berman, All That Is Solid Melts into Air, 19.

6 On citational networks, see Latour and Woolgar, Laboratory Life.

7 The CUHK group's study spanned the four-year period from 1995 to 1998. The HKU study used data for the period 1995–97. Meteorological data was obtained from the Hong Kong Observatory, and air pollutant concentrations were obtained from the Environmental Protection Department.

8 Wong et al., "Associations between Daily Mortalities from Respiratory and Cardiovascular Diseases and Air Pollution in Hong Kong, China," 31.

9 On the work of standards and standardization, see Bowker and Star, *Sorting Things Out*. Also see Dunn, "Standards and Person-Making in East Central Europe"; Lakoff, "Diagnostic Liquidity."

10 Fortun, "Biopolitics and the Informating of Environmentalism."

11 Jasanoff, "Taking Life."

12 Wong et al., "Effect of Air Pollution on Daily Mortality in Hong Kong," 339.

13 Cohen, *No Aging in India*, 90.

14 Wong et al., "Associations between Daily Mortalities from Respiratory and Cardiovascular Diseases and Air Pollution in Hong Kong, China," 33.

15 Homi Bhabha, following Derrida, elaborates the disturbing power of "being additional" in a postcolonial situation. "Coming 'after' the original, or in 'addition to' it, gives the supplementary question the advantage of introducing a sense of 'secondariness' or belatedness into the structure of the original demand. The supplementary strategy suggests that adding 'to' need not 'add up' but may disturb the calculation." See *The Location of Culture*, 155.

16 For an example of this form of argument, see McMichael et al., "Inappropriate Use of Daily Mortality Analyses to Estimate Longer-Term Mortality Effects of Air Pollution."

17 Jain, "Living in Prognosis."

18 Kuriyama, *The Expressiveness of the Body and the Divergence of Greek and Chinese Medicine*, 236.

19 Ibid., 245.

20 My thanks to an anonymous reviewer and to Rachel Prentice for encouraging me to draw this connection.

21 See, for instance, Feld and Basso, *Senses of Place*; Raffles, *In Amazonia*; Rodman, "Empowering Place."

22 Mong Kok's role in imaginations of Hong Kong is illuminated by the neighborhood's selection as a challenge in an American reality television show. Contestants were asked in other parts of Hong Kong to complete tasks such as lowering a shipping container with a crane or finding the tallest building in Central. In Mong Kok, however, their task was simply to find a certain tea shop where they would be asked to drink a bitter tea. Mong Kok's tightly packed and sporadically marked streets drove at least one contestant to tears.

23 Feld, "Waterfalls of Song."

24 See Harvey, *Justice, Nature and the Geography of Difference*.

25 Ehrlich, "A Breath of Fresh Poison."

26 For an analysis of state efforts in East and Southeast Asia to craft exceptional spaces attractive to foreign capital investment, see Ong, *Neoliberalism as Exception*. Ong adopts the term "ecologies" metaphorically to refer to the desired labor and financial conditions that are hoped to be conducive to a city's insertion in global trade circuits. I would merely add that in the pursuit of such desires, the ecologies

of landscapes, airscapes, and waterscapes can be equally subject to concern and government.

27 Rachel Stern provides a convincing and crucial argument for the recognition of air pollution as a social class issue in Hong Kong. She points out that, despite the fact that the lower classes suffer greater exposure to air pollution in their occupations and in their homes, the city's elites have generally set the clean-air political agenda. See "Hong Kong Haze."

28 Exemplary works charting the atmospheric differentiations of social class include Murphy, *Sick Building Syndrome and the Problem of Uncertainty*; Sze, *Noxious New York*.

29 For a discussion of the historical racialization of urban space in colonial Hong Kong, see Bremner and Lung, "Spaces of Exclusion."

30 Farquhar, *Appetites*, 46.

31 On the poetics of colonial concerns over pollution, contagion, and health, see Anderson, "Excremental Colonialism."

32 Quoted in Bremner and Lung, "Spaces of Exclusion," 244.

33 The following draws heavily upon material presented on the website of the United Kingdom's National Physical Laboratory, "What Is the History of Weighing? (FAQ—Mass and Density)," October 8, 2007, http://www.npl.co.uk/.

34 See Latour, *Pandora's Hope*.

35 The objective was 30,000 µg/m³ both in 2002, when the first version of this chapter was drafted, and again in 2007 when it was revised. See Environmental Protection Department, Government of Hong Kong, "Air Quality Objectives," December 5, 2006, http://www.epd-asg.gov.hk/english/backgd/hkaqo.php.

36 The World Health Organization's acceptable NO_2 annual mean is 40 µg/m³; Hong Kong's target is 80 µg/m³. WHO, "Air Quality and Health," Media Centre fact sheet N313, 2008, http://www.who.int/en/.

37 The World Health Organization's target for PM_{10} is 50 µg/m³; for $PM_{2.5}$ it is 25 µg/m³. Ibid.

38 For instance, former Hong Kong Legislative Council member Christine Loh made air quality and the state of Victoria Harbour central issues of her tenure in LegCo. Both in and out of office, she has consistently figured environmental issues as examples of how important it is for Hong Kong's government to heed the needs and voices of its public.

39 Dumit, "Prescription Maximization and the Accumulation of Surplus Health in the Pharmaceutical Industry."

40 Fortun, "Biopolitics and the Informating of Environmentalism."

41 Also relevant to this point is Adriana Petryna's account of Ukrainians' struggles after the Chernobyl accident to substantiate their debilitations as radiation-caused ailments in order to receive state aid while confronted with narrow and fluctuating definitions of radiation sickness. See *Life Exposed*.

42 See also Badiou, *Saint Paul*; Balibar, "Ambiguous Universality."

43 Laclau, "Identity and Hegemony," 58.

44 Butler, "Competing Universalities," 166.

45 On the distinction between the *a priori* and *a posteriori*, Butler remarks: "We might read the state of debate in which the *a priori* is consistently counterposed to the *a posteriori* as a symptom to be read, one that suggests something about the foreclosure of the conceptual field, its restriction to tired binary oppositions, one that is ready for a new opening" ("Dynamic Conclusions," 274). This is an argument that strict distinctions between the *a priori* and *a posteriori* signal a devolution of discussion rather than an elevated state, an argument that such oppositions indicate a foreclosed and unfruitful practice of theorizing. In the context of the exchange with Laclau and Žižek in which Butler makes this argument, this can be read as a claim that theory would benefit from an infusion of historical materiality. One should not mistake Butler's argument, however, for a call for empiricism against theory. This would repeat the same mistake of strict distinction. I take her statement, instead, as an invitation to think about theory's poetics.

46 Bernstein, *A Poetics*, 150–51.

47 Poetry is an act of creation for Bernstein, a hopeful writing with ambition to forge something new: "Poetry is aversion of conformity in the pursuit of new forms, or can be. By form I mean ways of putting things together, or stripping them apart, I mean ways of accounting for what weighs upon any one of us, or that poetry tosses up into an imaginary air like so many swans flying out of a magician's depthless black hat so that suddenly, like when the sky all at once turns white or purple or day-glo blue, we breathe more deeply" (1). The pursuit of new forms and the quest for new arrangements of things so that the skies change color around us, so we all may breathe more deeply—such a poetics strikes me as in tune with the aims of our most critical and ambitious theoretico-political projects.

Abbas, Ackbar. *Hong Kong: Culture and the Politics of Disappearance*. Minneapolis: University of Minnesota Press, 1997.

Abraham, Itty. *The Making of the Indian Atomic Bomb: Science, Secrecy and the Postcolonial State*. London: Zed Books, 1998.

Adams, Vincanne, and Stacy Leigh Pigg, eds. *Sex in Development: Science, Sexuality, and Morality in Global Perspective*. Durham, N.C.: Duke University Press, 2005.

Adler, Judith. "Cultivating Wilderness: Environmentalism and Legacies of Early Christian Asceticism." *Comparative Studies in Society and History* 48, no. 1 (2006), 4–37.

Agarwal, Anil, and Sunita Narain. *Global Warming in an Unequal World: A Case of Environmental Colonialism*. New Delhi: Centre for Science and Environment, 1991.

Agrawal, Arun. *Environmentality: Technologies of Government and the Making of Subjects*. Durham, N.C.: Duke University Press, 2005.

Anderson, Benedict. *The Spectre of Comparisons: Nationalism, Southeast Asia and the World*. London: Verso, 1998.

Anderson, H. Ross, Antonio Ponce de Leon, J. Martin Bland, Jonathan S. Bower, and David P. Strachan. "Air Pollution and Daily Mortality in London: 1987–92." *British Medical Journal* 312, no. 7032 (1996), 665–69.

Anderson, Warwick. "Excremental Colonialism: Public Health and the Poetics of Pollution." *Critical Inquiry* 21, no. 3 (1995), 640–70.

Badiou, Alain. *Saint Paul: The Foundation of Universalism*. Translated by Ray Brassier. Stanford, Calif.: Stanford University Press, 2003.

Balibar, Etienne. "Ambiguous Universality." *Differences* 7, no. 1 (1995), 48–74.

Bauman, Richard. "Mediational Performance, Traditionalization, and the Authorization of Discourse." *Verbal Art across Cultures: The Aesthetics and Proto-Aesthetics of Com-*

munication, edited by Hubert Knoblauch and Helga Kotthoff, 91–115. Tübingen, Germany: Gunter Narr Verlag, 2001.

———. *A World of Others' Words: Cross-Cultural Perspectives on Intertextuality*. Malden, Mass.: Blackwell, 2004.

Berman, Marshall. *All That Is Solid Melts into Air: The Experience of Modernity*. New York: Viking Penguin, 1988.

Bernstein, Charles. *A Poetics*. Cambridge, Mass.: Harvard University Press, 1992.

Bhabha, Homi K. *The Location of Culture*. New York: Routledge, 1994.

Boas, Franz. "The Limitations of the Comparative Method of Anthropology." *Science* 4, no. 103 (1896), 901–8.

Boellstorff, Tom. "Dubbing Culture: Indonesian Gay and Lesbi Subjectivities and Ethnography in an Already Globalized World." *American Ethnologist* 30, no. 2 (2003), 225–42.

Bowker, Geoffrey C., and Susan Leigh Star. *Sorting Things Out: Classification and Its Consequences*. Cambridge, Mass.: MIT Press, 1999.

Braun, Bruce. *The Intemperate Rainforest: Nature, Culture, and Power on Canada's West Coast*. Minneapolis: University of Minnesota Press, 2002.

Bremner, G. Alex, and David P. Y. Lung. "Spaces of Exclusion: The Significance of Cultural Identity in the Formation of European Residential Districts in British Hong Kong, 1877–1904." *Environment and Planning D* 21, no. 2 (2003), 223–52.

Briggs, Charles L. "Theorizing Modernity Conspiratorially: Science, Scale, and the Political Economy of Public Discourse Inexplanations of a Cholera Epidemic." *American Ethnologist* 31, no. 2 (2004), 164–87.

Brosius, J. Peter. "Negotiating Citizenship in a Commodified Landscape: The Case of Penan Hunter-Gatherers in Sarawak, East Malaysia." Social Science Research Council conference on "Cultural Citizenship in Southeast Asia," Honolulu, May 2–4, 1993.

Brown, Jacqueline Nassy. *Dropping Anchor, Setting Sail: Geographies of Race in Black Liverpool*. Princeton, N.J.: Princeton University Press, 2005.

Brown, Nicholas. "$\{\varnothing,\$\} \in \{\$\}$?: Or, Alain Badiou and Slavoj Žižek, Waiting for Something to Happen." *CR: The New Centennial Review* 4, no. 3 (2004), 289–319.

Brush, Stephen B., and Doreen Stabinsky. *Valuing Local Knowledge: Indigenous People and Intellectual Property Rights*. Washington, D.C.: Island Press, 1996.

Budgen, Sebastian, Eustache Kouvélakis, and Slavoj Žižek. *Lenin Reloaded: Toward a Politics of Truth*. Durham, N.C.: Duke University Press, 2007.

Butler, Judith. *Bodies That Matter: On the Discursive Limits Of "Sex"*. New York: Routledge, 1993.

———. "Competing Universalities." *Contingency, Hegemony, Universality: Contemporary Dialogues on the Left*, edited by Judith Butler, Ernesto Laclau, and Slavoj Žižek, 136–81. London: Verso, 2000.

———. "Dynamic Conclusions." *Contingency, Hegemony, Universality: Contemporary Dia-*

logues on the Left, edited by Judith Butler, Ernesto Laclau, and Slavoj Žižek, 263–80. London: Verso, 2000.

Butler, Judith, Ernesto Laclau, and Slavoj Žižek. Contingency, Hegemony, Universality: Contemporary Dialogues on the Left. London: Verso, 2000.

Callon, Michel. "Some Elements of a Sociology of Translation: Domestication of the Scallops and the Fishermen of St Brieuc Bay." Power, Action and Belief: A New Sociology of Knowledge? edited by John Law, 196–233. London: Routledge, 1986.

Chan, Selina Ching. "Negotiating Tradition: Customary Succession in the New Territories of Hong Kong." Hong Kong: The Anthropology of a Chinese Metropolis, edited by Grant Evans and Maria Tam Siu-Mi, 151–73. Honolulu: University of Hawai'i Press, 1997.

———. "Politicizing Tradition." Ethnology 37, no. 1 (1998), 39–54.

Cheah, Pheng. "Introduction Part II: The Cosmopolitical—Today." Cosmopolitics: Thinking and Feeling beyond the Nation, edited by Pheng Cheah and Bruce Robbins, 20–41. Minneapolis: University of Minnesota Press, 1998.

———. Spectral Nationality: Passages of Freedom from Kant to Postcolonial Literatures of Liberation. New York: Columbia University Press, 2003.

Cheah, Pheng, and Jonathan D. Culler, eds. Grounds of Comparison: Around the Work of Benedict Anderson. New York: Routledge, 2003.

Cheah, Pheng, and Bruce Robbins, eds. Cosmopolitics: Thinking and Feeling Beyond the Nation. Minneapolis: University of Minnesota Press, 1998.

Choy, Timothy K. "Politics by Example: An Ethnography of Environmental Emergences in Post-Colonial Hong Kong." Ph.D. Dissertation, University of California, Santa Cruz, 2003.

Chun, Allen John Uck. "Land Is to Live: A Study of the Concept of Tsu in a Hakka Chinese Village, New Territories, Hong Kong." Ph.D. Dissertation, University of Chicago, 1985.

———. Unstructuring Chinese Society: The Fictions of Colonial Practice and the Changing Realities of "Land" in the New Territories of Hong Kong. Amsterdam: Harwood Academic, 2000.

Clarke, S. C., A. P. Jackson, and J. Neff. "Development of a Risk Assessment Methodology for Evaluating Potential Impacts Associated with Contaminated Mud Disposal in the Marine Environment." Chemosphere 41 (2000), 69–76.

Clifford, James. "Mixed Feelings." Cosmopolitics: Thinking and Feeling Beyond the Nation, edited by Pheng Cheah and Bruce Robbins, 362–70. Minneapolis: University of Minnesota Press, 1998.

Cohen, Lawrence. No Aging in India: Alzheimer's, the Bad Family, and Other Modern Things. Berkeley: University of California Press, 1998.

Collins, Harry M., and Trevor J. Pinch. The Golem: What Everyone Should Know About Science. Cambridge: Cambridge University Press, 1993.

Constable, Nicole. Maid to Order in Hong Kong: Stories of Filipina Workers. Ithaca, N.Y.: Cornell University Press, 1997.

Cooke, Bill, and Uma Kothari. *Participation: The New Tyranny?* London: Zed Books, 2001.

Covey, J. G. "Accountability and Effectiveness in NGO Policy Alliances." *Journal of International Development* 7 (1995), 857–68.

Daston, Lorraine. "Objectivity and the Escape from Perspective." *Social Studies of Science* 22, no. 4 (1992), 597–618.

Di Chiro, Giovanna. "Defining Environmental Justice: Women's Voices and Grassroots Politics." *Socialist Review* 22, no. 4 (1992), 93–130.

Dumit, Joseph. *Picturing Personhood: Brain Scans and Biomedical Identity.* Princeton, N.J.: Princeton University Press, 2003.

———. "Prescription Maximization and the Accumulation of Surplus Health in the Pharmaceutical Industry: The BioMarx Experiment." *Lively Capital: Biotechnologies, Ethics, and Governance in Global Markets*, edited by Kaushik Sunder Rajan. Durham, N.C.: Duke University Press, forthcoming.

Dunn, Elizabeth C. "Standards and Person-Making in East Central Europe." *Global Assemblages: Technology, Politics, and Ethics as Anthropological Problems*, edited by Aihwa Ong and Stephen J. Collier, 173–93. Malden, Mass.: Blackwell, 2005.

Ehrlich, Jennifer. "A Breath of Fresh Poison." *South China Morning Post*, August 14, 2000.

Endacott, G. B. *A History of Hong Kong.* Revised edition. Hong Kong: Oxford University Press, 1973.

Escobar, Arturo. "After Nature: Steps to an Anti-Essentialist Political Ecology." *Current Anthropology* 40, no. 1 (1999), 1–30.

Farquhar, Judith. *Appetites: Food and Sex in Post-socialist China.* Durham, N.C.: Duke University Press, 2002.

Feld, Steven. "Waterfalls of Song: An Acoustemology of Place Resounding in Bosavi, Papua New Guinea." *Senses of Place*, edited by Steven Feld and Keith H. Basso, 91–135. Santa Fe: School of American Research Press, 1996.

Feld, Steven, and Keith H. Basso, eds. *Senses of Place.* Santa Fe: School of American Research Press, 1996.

Fleck, Ludwik. *Genesis and Development of a Scientific Fact.* Chicago: University of Chicago Press, 1979.

Forsyth, Tim. "Social Movements and Environmental Democratization in Thailand." *Earthly Politics: Local and Global in Environmental Governance*, edited by Sheila Jasanoff and Marybeth Long Martello, 195–215. Cambridge, Mass.: MIT Press, 2004.

Fortun, Kim. *Advocacy after Bhopal: Environmentalism, Disaster, New Global Orders.* Chicago: University of Chicago Press, 2001.

———. "The Bhopal Disaster: Advocacy and Expertise." *Science as Culture* 7, no. 2 (1998), 193–216.

———. "Biopolitics and the Informating of Environmentalism." *Lively Capital: Biotechnologies, Ethics, and Governance in Global Markets*, edited by Kaushik Sunder Rajan. Durham, N.C.: Duke University Press, forthcoming.

Foucault, Michel. *Discipline and Punish: The Birth of the Prison.* Translated by Alan Sheridan. New York: Vintage Books, 1979.

Geertz, Clifford. "Local Knowledge: Fact and Law in Comparative Perspective." *Local Knowledge: Further Essays in Interpretive Anthropology,* 167–234. New York: Basic Books, 1983.

Gibson-Graham, J. K. *The End of Capitalism (as We Knew It): A Feminist Critique of Political Economy.* Cambridge, Mass.: Blackwell, 1996.

Gillmor, Dan. "Proposed Security Law Putting Liberties at Risk in Hong Kong." *San Jose Mercury News* (Calif.), December 15, 2002, 1F.

Golan, Tal. *Laws of Men and Laws of Nature: The History of Scientific Expert Testimony in England and America.* Cambridge, Mass.: Harvard University Press, 2004.

Golley, Frank Benjamin. *A History of the Ecosystem Concept in Ecology: More Than the Sum of the Parts.* New Haven, Conn.: Yale University Press, 1993.

Gould, Kenneth Alan, Allan Schnaiberg, and Adam S. Weinberg. *Local Environmental Struggles: Citizen Activism in the Treadmill of Production.* Cambridge: Cambridge University Press, 1996.

Gramsci, Antonio. *Selections from the Prison Notebooks.* Edited and translated by Quintin Hoare and Geoffrey Nowell Smith. London: Lawrence and Wishart, 1971.

Grove, Richard H. *Green Imperialism: Colonial Expansion, Tropical Island Edens and the Origins of Environmentalism, 1600–1860.* Cambridge: Cambridge University Press, 1995.

Guldin, Gregory Eliyu. "Hong Kong Ethnicity: Of Folk Models and Change." *Hong Kong: The Anthropology of a Chinese Metropolis,* edited by Grant Evans and Maria Tam Siu-Mi, 25–50. Honolulu: University of Hawai'i Press, 1997.

Gupta, Akhil. *Postcolonial Developments: Agriculture in the Making of Modern India.* Durham, N.C.: Duke University Press, 1998.

Gupta, Akhil, and James Ferguson, eds. *Culture, Power, Place: Explorations in Critical Anthropology.* Durham, N.C.: Duke University Press, 1997.

Hall, Stuart, and Lawrence Grossberg. "On Postmodernism and Articulation: An Interview with Stuart Hall." *Stuart Hall: Critical Dialogues in Cultural Studies,* edited by David Morley and Kuan-Hsing Chen, 151–73. London: Routledge, 1996.

Haraway, Donna J. *Modest_Witness@Second_Millennium.FemaleMan©_Meets_OncoMouse™.* New York: Routledge, 1997.

———. *Primate Visions: Gender, Race, and Nature in the World of Modern Science.* New York: Routledge, 1989.

———. "Situated Knowledges: The Science Question in Feminism and the Privilege of Partial Perspective." *Simians, Cyborgs, and Women: The Reinvention of Nature,* 183–201. New York: Routledge, 1991.

Harding, Susan F. *The Book of Jerry Falwell: Fundamentalist Language and Politics.* Princeton, N.J.: Princeton University Press, 2000.

———. "Convicted by the Holy Spirit: The Rhetoric of Fundamental Baptist Conversion." *American Ethnologist* 14, no. 1 (1987), 167–81.

Harper, Richard. "The Social Organization of the IMF's Mission Work: An Examination of International Auditing." *Audit Cultures: Anthropological Studies in Accountability, Ethics, and the Academy*, edited by Marilyn Strathern, 21–53. London: Routledge, 2000.

Harvey, David. *A Brief History of Neoliberalism*. Oxford: Oxford University Press, 2005.

———. *The Condition of Postmodernity: An Enquiry into the Origins of Cultural Change*. Oxford: Blackwell, 1989.

———. *Justice, Nature and the Geography of Difference*. Cambridge, Mass.: Blackwell, 1996.

Hayden, Cori. *When Nature Goes Public: The Making and Unmaking of Bioprospecting in Mexico*. Princeton, N.J.: Princeton University Press, 2003.

Helmreich, Stefan. *Silicon Second Nature: Culturing Artificial Life in a Digital World*. Berkeley: University of California Press, 1998.

Ho, Karen. "Situating Global Capitalisms: A View from Wall Street Investment Banks." *Cultural Anthropology* 20, no. 1 (2005), 68–96.

Ho, Louise. "Storm." *New Ends, Old Beginnings*, 54. Hong Kong: Asia 2000, 1997.

Hong Kong Research Grants Council. *Annual Report 1999*. Hong Kong, China: Hong Kong Research Grants Council, 2000.

Hong Yun-Chul, Jong-Han Leem, Eun-Hee Ha, and David C. Christiani. "PM_{10} Exposure, Gaseous Pollutants, and Daily Mortality in Inchon, South Korea." *Environmental Health Perspectives* 54 (1999), 108–16.

Hu Shiu-ying. *The Genera of Orchidaceae in Hong Kong*. Hong Kong: Chinese University of Hong Kong, 1977.

Hu Shiu-ying, and Gloria Barretto. "New Species and Varieties of Orchidaceae in Hong Kong." *Chung Chi Journal* (1975), 1–34.

Ingold, Tim. "Globes and Spheres: The Topology of Environmentalism." *Environmentalism: The View from Anthropology*, edited by Kay Milton, 31–42. London: Routledge, 1993.

Ivy, Marilyn. *Discourses of the Vanishing: Modernity, Phantasm, Japan*. Chicago: University of Chicago Press, 1995.

Jain, Sarah Lochlann. "Living in Prognosis: Toward an Elegiac Politics." *Representations* 98, no. 1 (2007), 77–92.

Jameson, Fredric. "Postmodernism, or the Cultural Logic of Late Capitalism." *New Left Review* 146 (1984), 53–92.

Jasanoff, Sheila. *The Fifth Branch: Science Advisers as Policymakers*. Cambridge, Mass.: Harvard University Press, 1990.

———. "Heaven and Earth: The Politics of Earthly Images." *Earthly Politics: Local and Global in Environmental Governance*, edited by Sheila Jasanoff and Marybeth Long Martello, 31–52. Cambridge, Mass.: MIT Press, 2004.

———. "(No?) Accounting for Expertise." *Science and Public Policy* 30, no. 3 (2003), 157–63.

———. "Taking Life: Private Rights in Public Nature." *Lively Capital: Biotechnologies,*

Ethics, and Governance in Global Markets, edited by Kaushik Sunder Rajan. Durham, N.C.: Duke University Press, forthcoming.

Jasanoff, Sheila, and Marybeth Long Martello, eds. Earthly Politics: Local and Global in Environmental Governance. Cambridge, Mass.: MIT Press, 2004.

Kant, Immanuel. Critique of Pure Reason. Translated by Norman Kemp Smith. New York: St. Martin's Press, 1965.

———. "Idea for a Universal History with a Cosmopolitan Intent." Perpetual Peace, and Other Essays on Politics, History, and Morals, edited by Ted Humphrey, 29–40. Indianapolis: Hackett, 1983.

Kaplan, Caren. Questions of Travel: Postmodern Discourses of Displacement. Durham, N.C.: Duke University Press, 1996.

Kirksey, S. Eben. "Foam Frogs and Eco-Tractors: Voices from a Neotropical Swamp." Paper presented at the annual meeting of the American Anthropological Association, San Jose, Calif., November 15–17, 2006.

Kosek, Jake. Understories: The Political Life of Forests in Northern New Mexico. Durham, N.C.: Duke University Press, 2006.

Kuletz, Valerie. The Tainted Desert: Environmental Ruin in the American West. New York: Routledge, 1998.

Kuriyama, Shigehisa. The Expressiveness of the Body and the Divergence of Greek and Chinese Medicine. New York: Zone Books, 1999.

Kurtz, Donald V. "Hegemony and Anthropology: Gramsci, Exegeses, Reinterpretations." Critique of Anthropology 16, no. 2 (1996), 103–35.

Kurzman, Steven. "Embodiment and Ability: Amputees and Prostheses in America." Ph.D. Dissertation, University of California, Santa Cruz, 2003.

Kwa, Chunglin. "Romantic and Baroque Conceptions of Complex Wholes in the Sciences." Complexities: Social Studies of Knowledge Practices, edited by John Law and Annemarie Mol, 23–52. Durham, N.C.: Duke University Press, 2002.

Laclau, Ernesto. "Identity and Hegemony: The Role of Universality in the Constitution of Political Logics." Contingency, Hegemony, Universality: Contemporary Dialogues on the Left, edited by Judith Butler, Ernesto Laclau, and Slavoj Žižek, 44–89. London: Verso, 2000.

———. Politics and Ideology in Marxist Theory. London: Verso, 1979.

Laclau, Ernesto, and Chantal Mouffe. Hegemony and Socialist Strategy: Towards a Radical Democratic Politics. Translated by Winston Moore and Paul Cammack. London: Verso, 1985.

Lai On Kwok. "Greening of Hong Kong?—Forms of Manifestation of Environmental Movements." The Dynamics of Social Movement in Hong Kong, edited by Stephen Wing Kai Chiu and Tai Lok Lui, 259–95. Hong Kong: Hong Kong University Press, 2000.

Lakoff, Andrew. "Diagnostic Liquidity: Mental Illness and the Global Trade in DNA." Theory and Society 34, no. 1 (2005), 63–92.

Langwick, Stacey A. "Articulate(d) Bodies: Traditional Medicine in a Tanzanian Hospital." *American Ethnologist* 35, no. 3 (2008), 428–39.

———. *Bodies, Politics and African Healing: The Matter of Maladies in Tanzania.* Bloomington: Indiana University Press, 2011.

———. "Devils, Parasites, and Fierce Needles: Healing and the Politics of Translation in Southern Tanzania." *Science, Technology and Human Values* 32, no. 1 (2007), 88–117.

Latour, Bruno. *Aramis, or the Love of Technology.* Translated by Catherine Porter. Cambridge, Mass.: Harvard University Press, 1996.

———. *Pandora's Hope: Essays on the Reality of Science Studies.* Cambridge, Mass.: Harvard University Press, 1999.

———. *Science in Action: How to Follow Scientists and Engineers through Society.* Cambridge, Mass.: Harvard University Press, 1987.

———. *We Have Never Been Modern.* Translated by Catherine Porter. Cambridge, Mass.: Harvard University Press, 1993.

Latour, Bruno, and Steve Woolgar. *Laboratory Life: The Construction of Scientific Facts.* Princeton, N.J.: Princeton University Press, 1986.

Law, Wing-sang. "Northbound Colonialism: A Politics of Post-PC Hong Kong." *Positions* 8, no. 1 (2000), 201–33.

Lee Chin-chuan, Joseph Man Chan, Clement Y. K. So, and Zhongdan Pan. *Global Media Spectacle: News War over Hong Kong.* Albany: State University of New York Press, 2002.

Lenoir, Timothy. *The Strategy of Life: Teleology and Mechanics in Nineteenth-Century German Biology.* Chicago: University of Chicago Press, 1989.

Li, Tania Murray. "Articulating Indigenous Identity in Indonesia: Resource Politics and the Tribal Slot." *Comparative Studies in Society and History* 42, no. 1 (2000), 149–73.

Liu, J. H., and Peter Hills. "Environmental Planning, Biodiversity and the Development Process: The Case of Hong Kong's Chinese White Dolphins." *Journal of Environmental Management* 50 (1997), 351–67.

Liu, Lydia He. *Translingual Practice: Literature, National Culture, and Translated Modernity—China, 1900–1937.* Stanford, Calif.: Stanford University Press, 1995.

Lock, Margaret M. *Twice Dead: Organ Transplants and the Reinvention of Death.* Berkeley: University of California Press, 2002.

Lorimer, Jamie. "Nonhuman Charisma," *Environment & Planning D* 25, no. 5 (2007), 911–32.

Low, Setha M., and Denise Lawrence-Zúñiga. *The Anthropology of Space and Place: Locating Culture.* Malden, Mass: Blackwell, 2003.

Lowe, Celia. "Making the Monkey: How the Togean Macaque Went from 'New Form' to 'Endemic Species' in Indonesians' Conservation Biology." *Cultural Anthropology* 19, no. 4 (2004), 491–516.

———. *Wild Profusion: Biodiversity Conservation in an Indonesian Archipelago.* Princeton, N.J.: Princeton University Press, 2006.

Ma, Eric Kit-wai. "Re-Advertising Hong Kong: Nostalgia Industry and Popular History." *Positions* 9, no. 1 (2001), 131–59.

Mahmood, Saba. *Politics of Piety: The Islamic Revival and the Feminist Subject*. Princeton, N.J.: Princeton University Press, 2005.

Man Si-wai. "The Environment." *From Colony to SAR: Hong Kong's Challenges Ahead*, edited by Joseph Y. S. Cheng and Sonny S. H. Lo, 319–56. Hong Kong: Chinese University Press, 1995.

Man Si-wai, and Lo Sze-ping, eds. "Cultural Identities and Cultural Politics: Colonial and Postcolonial Imaginations in Hong Kong." Special issue, *Chinese Sociology and Anthropology* 30, no. 3 (1998).

Martin, Emily. *Flexible Bodies: Tracing Immunity in American Popular Culture from the Days of Polio to the Age of AIDS*. Boston: Beacon Press, 1994.

———. *The Woman in the Body: A Cultural Analysis of Reproduction*. Boston: Beacon Press, 1992.

Marx, Karl. "Manifesto of the Communist Party." *The Marx-Engels Reader*, edited by Robert Tucker, 469–500. New York: W. W. Norton, 1978.

———. "Speech at the Anniversary of the *People's Paper*." *The Marx-Engels Reader*, edited by Robert Tucker, 577–78. New York: W. W. Norton, 1978.

Mathews, Gordon. "Hèunggóngyàhn: On the Past, Present and Future of Hong Kong Identity." *Bulletin of Concerned Asian Scholars* 29, no. 3 (1997), 3–13.

Maurer, Bill. "Anthropological and Accounting Knowledge in Islamic Banking and Finance: Rethinking Critical Accounts." *Journal of the Royal Anthropological Institute* 8, no. 4 (2002), 645–67.

———. *Mutual Life, Limited: Islamic Banking, Alternative Currencies, Lateral Reason*. Princeton, N.J.: Princeton University Press, 2005.

McMichael, A. J., H. R. Anderson, B. Brunekreef, and A. J. Cohen. "Inappropriate Use of Daily Mortality Analyses to Estimate Longer-Term Mortality Effects of Air Pollution." *International Journal of Epidemiology* 27, no. 3 (1998), 450–53.

Merry, Sally Engle, and Rachel Stern. "The Female Inheritance Movement in Hong Kong: Theorizing the Local/Global Interface." *Current Anthropology* 46, no. 3 (2005), 387–408.

Mitchell, Katharyne. "Transnational Subjects: Constituting the Cultural Citizen in the Era of Pacific Rim Capital." *Ungrounded Empires: The Cultural Politics of Modern Chinese Transnationalism*, edited by Aihwa Ong and Donald Nonini, 228–56. New York: Routledge, 1997.

Mitchell, Timothy. *Rule of Experts: Egypt, Techno-Politics, Modernity*. Berkeley: University of California Press, 2002.

Miyazaki, Hirokazu. "The Temporalities of the Market." *American Anthropologist* 105, no. 2 (2003), 255–65.

Miyazaki, Hirokazu, and Annelise Riles. "Failure as an Endpoint." *Global Assemblages:*

Technology, Politics, and Ethics as Anthropological Problems, edited by Aihwa Ong and Stephen J. Collier, 320–31. Malden, Mass.: Blackwell, 2005.

Mohanty, Chandra Talpade, Ann Russo, and Lourdes Torres. Third World Women and the Politics of Feminism. Bloomington: Indiana University Press, 1991.

Moolgavkar, Suresh H., George E. Luebeck, Thomas A. Hall, and Elizabeth L. Anderson. "Air Pollution and Daily Mortality in Philadelphia." Epidemiology 6, no. 5 (1995), 476–84.

Moore, Donald. "Contesting Terrain in Zimbabwe's Eastern Highlands: Political Ecology and Peasant Resource Struggles." Economic Geography 69, no. 3 (1993), 380–401.

———. "The Crucible of Cultural Politics: Reworking 'Development' in Zimbabwe's Eastern Highlands." American Ethnologist 26, no. 3 (1999), 654–89.

Moore, Donald S., Jake Kosek, and Anand Pandian. "The Cultural Politics of Race and Nature: Terrains of Power and Practice." Race, Nature, and the Politics of Difference, edited by Donald S. Moore, Jake Kosek, and Anand Pandian, 1–70. Durham, N.C.: Duke University Press, 2003.

Murphy, Michelle. Sick Building Syndrome and the Problem of Uncertainty: Environmental Politics, Technoscience, and Women Workers. Durham, N.C.: Duke University Press, 2006.

Ong, Aihwa. "Cyberpublics and Diaspora Politics among Transnational Chinese." Interventions 5, no. 1 (2003), 82–100.

———. Flexible Citizenship: The Cultural Logics of Transnationality. Durham, N.C.: Duke University Press, 1999.

———. Neoliberalism as Exception: Mutations in Citizenship and Sovereignty. Durham, N.C.: Duke University Press, 2006.

Pellow, David Naguib. Garbage Wars: The Struggle for Environmental Justice in Chicago. Cambridge, Mass.: MIT Press, 2002.

Petryna, Adriana. Life Exposed: Biological Citizens after Chernobyl. Princeton, N.J.: Princeton University Press, 2002.

Pigg, Stacy Leigh. "Languages of Sex and AIDS in Nepal: Notes on the Social Production of Commensurability." Cultural Anthropology 16, no. 4 (2001), 481–541.

Power, Michael. The Audit Society: Rituals of Verification. 2nd edition. New York: Oxford University Press, 1999.

Pratt, Mary Louise. "Alexander von Humboldt and the Reinvention of América." Imperial Eyes: Travel Writing and Transculturation, 111–43. New York: Routledge, 1992.

———. Imperial Eyes: Travel Writing and Transculturation. New York: Routledge, 1992.

———. "Science, Planetary Consciousness, Interiors." Imperial Eyes: Travel Writing and Transculturation, 15–37. New York: Routledge, 1992.

Rabinow, Paul. French DNA: Trouble in Purgatory. Chicago: University of Chicago Press, 1999.

———. Making PCR: A Story of Biotechnology. Chicago: University of Chicago Press, 1996.

Rafael, Vicente L. *Contracting Colonialism: Translation and Christian Conversion in Tagalog Society under Early Spanish Rule.* Ithaca, N.Y.: Cornell University Press, 1988.

Raffles, Hugh. *In Amazonia: A Natural History.* Princeton, N.J.: Princeton University Press, 2002.

Rapp, Rayna. *Testing Women, Testing the Fetus: The Social Impact of Amniocentesis in America.* New York: Routledge, 1999.

Reardon, Jenny. *Race to the Finish: Identity and Governance in an Age of Genomics.* Princeton, N.J.: Princeton University Press, 2004.

Riles, Annelise. *Collateral Knowledge: Legal Reasoning in the Global Financial Markets.* Chicago: University of Chicago Press, 2011.

———. *The Network Inside Out.* Ann Arbor: University of Michigan Press, 2000.

Rodman, Margaret. "Empowering Place: Multilocality and Multivocality." *American Anthropologist* 94 (1992), 640–56.

Rofel, Lisa. *Other Modernities: Gendered Yearnings in China after Socialism.* Berkeley: University of California Press, 1999.

Rosaldo, Renato. *Culture and Truth: The Remaking of Social Analysis.* Boston: Beacon Press, 1993.

Satsuka, Shiho. "Traveling Nature, Imagining the Globe: Japanese Tourism in the Canadian Rockies." Ph.D. Dissertation, University of California, Santa Cruz, 2004.

Sawyer, Suzana. *Crude Chronicles: Indigenous Politics, Multinational Oil, and Neoliberalism in Ecuador.* Durham, N.C.: Duke University Press, 2004.

Sawyer, Suzana, and Arun Agrawal. "Environmental Orientalisms." *Cultural Critique* 45 (2000), 71–108.

Schwartz, J. "Particulate Air Pollution and Daily Mortality in Detroit." *Environmental Research* 56, no. 2 (1991), 204–13.

Schwartz, Joel, and Douglas W. Dockery. "Particulate Air Pollution and Daily Mortality in Steubenville, Ohio." *American Journal of Epidemiology* 135, no. 1 (1992), 12–19.

Scott, James C. *Domination and the Arts of Resistance: Hidden Transcripts.* New Haven, Conn.: Yale University Press, 1990.

———. *Weapons of the Weak: Everyday Forms of Peasant Resistance.* New Haven, Conn.: Yale University Press, 1985.

Shapin, Steven, and Simon Schaffer. *Leviathan and the Air-Pump: Hobbes, Boyle, and the Experimental Life.* Princeton, N.J.: Princeton University Press, 1989.

Silverstein, Michael. "Shifters, Linguistic Categories, and Cultural Description." *Meaning in Anthropology,* edited by Keith H. Basso and Henry A. Selby, 11–56. Albuquerque: University of New Mexico Press, 1976.

Star, Susan Leigh, and James R. Griesemer. "Institutional Ecology, 'Translations' and Boundary Objects: Amateurs and Professionals in Berkeley's Museum of Vertebrate Zoology, 1907–39." *Social Studies of Science* 19, no. 3 (1989), 387–420.

Stengers, Isabelle. *Cosmopolitiques.* 2 vols. Paris: La Découverte, 2003.

Stern, Rachel. "Hong Kong Haze: Air Pollution as a Social Class Issue." *Asian Survey* 43, no. 5 (2003), 780–800.

Stewart, Kathleen. "Nostalgia—A Polemic." *Cultural Anthropology* 3, no. 3 (1988), 227–41.

———. *A Space on the Side of the Road: Cultural Poetics in an "Other" America*. Princeton, N.J.: Princeton University Press, 1996.

Strathern, Marilyn. *Audit Cultures: Anthropological Studies in Accountability, Ethics, and the Academy*. London: Routledge, 2000.

———. *Partial Connections*. Savage, Md.: Rowman and Littlefield, 1991.

Sturgeon, Noël. *Ecofeminist Natures: Race, Gender, Feminist Theory, and Political Action*. New York: Routledge, 1997.

Sun, Mei. "The Allopolyploid Origin of *Spiranthes hongkongensis* (Orchidaceae)." *American Journal of Botany* 83, no. 2 (1996), 252–60.

Sunder Rajan, Kaushik. *Biocapital: The Constitution of Postgenomic Life*. Durham, N.C.: Duke University Press, 2006.

Sunyer, J., J. Castellsagué, M. Sáez, A. Tobias, J. M. Antó. "Air Pollution and Mortality in Barcelona." *Journal of Epidemiology and Community Health* 50, supplement 1 (1996), S76–S80.

Szasz, Andrew. *Ecopopulism: Toxic Waste and the Movement for Environmental Justice*. Minneapolis: University of Minnesota Press, 1994.

Sze, Julie. *Noxious New York: The Racial Politics of Urban Health and Environmental Justice*. Cambridge, Mass.: MIT Press, 2007.

Tam, Siumi Maria. "Eating Metropolitaneity: Hong Kong Identity in Yumcha." *Australian Journal of Anthropology* 8, no. 3 (1997), 291–306.

———. "*Heunggongyan* Forever: Immigrant Life and Hong Kong Style *Yumcha* in Australia." *The Globalization of Chinese Food*, edited by David Y. H. Wu and Sidney C. H. Cheung, 131–51. Honolulu: University of Hawai'i Press, 2002.

Taylor, Bron. "The Tributaries of Radical Environmentalism." *Journal for the Study of Radicalism* 2, no. 1 (2008), 27–61.

Toe Yuen. *Mak Dau Goo Si (My life as McDull)*. Hong Kong: EDKO Film, 2001.

Touloumi, G., E. Samoli, and K. Katsouyanni. "Daily Mortality and 'Winter Type' Air Pollution in Athens, Greece: A Time Series Analysis within the APHEA Project." *Journal of Epidemiology and Community Health* 50, supplement 1 (1996), S47–S51.

Traweek, Sharon. *Beamtimes and Lifetimes: The World of High Energy Physicists*. Cambridge, Mass: Harvard University Press, 1992.

Tsing, Anna. *Friction: An Ethnography of Global Connection*. Princeton, N.J.: Princeton University Press, 2004.

———. "The Global Situation." *Cultural Anthropology* 15, no. 3 (2000), 327–60.

———. "Inside the Economy of Appearances." *Public Culture* 12, no. 1 (2000), 115–42.

———. "Kinship and Science in the Genus Tricholoma." Paper presented at the annual meeting of the American Anthropological Association, San Francisco, November 19–23, 2008.

———. "Transitions as Translations." *Transitions, Environments, Translations: Feminisms in International Politics*, edited by Joan W. Scott, Cora Kaplan, and Debra Keates, 253–72. New York: Routledge, 1997.

Urban, Greg. *Metaculture: How Culture Moves through the World*. Minneapolis: University of Minnesota Press, 2001.

Watson, James L., ed. *Golden Arches East: McDonald's in East Asia*. Stanford, Calif.: Stanford University Press, 1997.

Weber, Max. *The Vocation Lectures*. Edited by David Owen and Tracy B. Strong. Translated by Rodney Livingstone. Indianapolis: Hackett, 2004.

Welker, Marina A. "'Corporate Security Begins in the Community': Mining, the Corporate Social Responsibility Industry, and Environmental Advocacy in Indonesia." *Cultural Anthropology* 24, no. 1 (2009), 142–79.

Wong, Chit-ming, Stefan Ma, Anthony Johnson Hedley, and Tai-hing Lam. "Effect of Air Pollution on Daily Mortality in Hong Kong." *Environmental Health Perspectives* 109, no. 4 (2001), 335–40.

Wong Kar-wai. *Chungking Express*. Hong Kong: Jet Tone Production, 1996.

Wong, T. W., W. S. Tam, T. S. Yu, and A. H. S. Wong. "Associations between Daily Mortalities from Respiratory and Cardiovascular Diseases and Air Pollution in Hong Kong, China." *Occupational and Environmental Medicine* 59, no. 1 (2002), 30–35.

Wong Wai King. *Tai O: Love Stories of the Fishing Village*. Hong Kong: Stepforward Multimedia Company, 2000.

Xi Xi. *Marvels of a Floating City and Other Stories: An Authorized Collection*. Translated by Eva Hung. Hong Kong: Research Centre for Translation, Chinese University of Hong Kong, 1997.

Xu, X. P., D. W. Dockery, and J. Gao. "Air Pollution and Daily Mortality in Residential Areas of Beijing, China." *Archives of Environmental Health* 49 (1994), 216–22.

Zhan, Mei. "Civet Cats, Fried Grasshoppers, and David Beckham's Pajamas: Unruly Bodies after SARS." *American Anthropologist* 107, no. 1 (2005), 31–42.

———. "Does It Take a Miracle? Negotiating Knowledges, Identities, and Communities in Traditional Chinese Medicine." *Cultural Anthropology* 16, no. 4 (2001), 453–80.

Žižek, Slavoj. *Repeating Lenin*. Zagreb: Bastard Books, 2001.

Jain, Sarah Lochlann, 151
Japan, authentic, 40
Jasanoff, Sheila, 149
Jiliu, 97–100

Kant, Immanuel, 63, 64
Kuriyama, Shigehisa, 153
Kurzman, Steven, 86

Laclau, Ernesto, 83
Landfills, 77
Land reclamation, 24
Land rights, 2–4, 16, 102
Lands Department (Hong Kong), 2
Langwick, Stacey, 90–91
Lantau Island, 7–9, 25, 36–37, 39, 46,
    51–52, 109. See also Tai O village
Lee, William, 124–33, 134
Legislative Council (LegCo), 85–86, 103
Lenoir, Timothy, 63
Li, Tania, 94
Local appropriateness: accountability and,
    88–89; environmentalism and, 17; en-
    vironmental politics and, 5, 11, 13, 104;
    local data and, 150; specificity and, 13;
    technologies and, 86–87, 104; univer-
    sal/particular and, 89, 104
Local knowledge, 80, 88, 104, 173 n. 4
Lung Kwu Tan village, 76, 77–79; collabo-
    ration with Greenpeace, 78–81, 89,
    94, 96, 103; collaboration with Green
    Power, 82
Lyle, Norman, 69–70

Ma, Eric, 40–41
Magritte, René, 139
Man Si-wai, 30–31
Maria. See Deng, Maria
Marx, Karl, 143–44, 145
Marxism, 20, 144
Mathews, Gordon, 31, 60
Mediational speech event, 91

Mong Kok district, 155, 156
Monsanto (company), 22
Mouffe, Chantal, 93

New Airport Projects Coordination Office
    (NAPCO), 24
New Territories, 8, 36, 101–2, 155. See also
    specific villages
Nongovernmental organizations (NGOs):
    collaboration with villages of, 5, 13–14,
    78–82, 89–90, 96, 103, 165; environ-
    mental, 1–2, 7–9, 13. See also specific orga-
    nizations
Nostalgia: anticipatory, 13, 28, 38–39,
    49–50; endangerment and, 28, 38,
    48–50, 62; immediacy of, 39; imperial-
    ist, 40; Lantau Island and, 36; salvag-
    ing, 49; subject of, 40–41; temporal
    modality of, 43, 48–49
Nuclear energy. See Antinuclear demon-
    strations

Ong, Aihwa, 134
On Kwok Lai, 10
Orchid studies, 53–59, 55, 65–66, 71;
    comparison and, 13, 56–58; specificity
    and, 56, 58
Organismic life, 64, 65, 70
Oscillation method, 15, 81

Paang uk. See Stilt houses
Particularity: of place, 112, 148; univer-
    sality and, 14, 79–81, 87–90, 95, 104–5,
    112, 143, 166–67
Pat Sin Leng Country Park, 2
Penny's Bay, 109
Pigg, Stacy Leigh, 90–91
Place: environmental politics and, 78; par-
    ticularity of, 112, 148; poetics of, 154;
    specificity of, 17–18, 148; translation
    and, 14. See also Space
Poetics: of air, 146, 155–57, 164–65; of

difference, 161; philosophy and, 168; of place, 154

Provisional Airport Authority, 28

Qi. See Hei

Radio Television Hong Kong, 33–34
Remembering-recording. See Geiluhk
Rio Declaration, 88
Rockey, John, 86–87
Rofel, Lisa, 41
Rosaldo, Renato, 40

Saltwater songs. See Haahmshuigo
Sha Lo Tung Development Company, 2
Sha Lo Tung environmental incident, 1, 2–4, 16–17
Shufuhk, 154, 157
Sino–British Joint Declaration (1984), 10, 27, 68
Site of Special Scientific Interest, 2–3, 17
Sousa chinensis, 5, 24. See also Dolphin protection
Sovereignty of Hong Kong, 5, 8, 40, 57, 62, 63; concept of nation and, 65; politics of conservation and, 70; specific life and, 67, 69
Space: coeval, 49–50; endangerment and, 27–28; loss and, 49. See also Place
Special Administration Region, 140
Speciation, 56, 60–61, 66, 67
Specificity: autonomy and, 70; in botany, 56, 58, 66–67; cultural, 62; endangerment and, 29–30; environmental politics and, 11, 12, 16; ethnographic research and, 17; freedom and, 69; local appropriateness and, 13; place-based, 17–18, 148; politics of, 5; rarity and, 66; specific life, 50, 67, 70–71
Spiranthes hongkongensis, 13, 54, 57, 59, 66. See also Orchid studies
Stewart, Kathleen, 41

Stilt houses: destruction of, 26, 43; rebuilding, 45, 55; in Tai O, 25–26, 37–38, 43–45, 170 n. 2
Sun, Mei, 57–60, 65–66, 67
Sun Yat-Sen, 60
Swire Institute of Marine Sciences (SWIMS), 28–30

Tai O Residents' Rights Committee, 45
Tai O village: atmosphere of, 138; fire destruction, 26, 43–45; indigenous village designation, 44, 45; loss and, 43–49; stilt houses, 25–26, 37–38, 43–45; tourism and, 25–26, 37–41, 155
Tam, Maria, 60–61, 62
Technologies, locally appropriate, 86–87, 104
TEK. See Traditional ecological knowledge (TEK)
Temporality: endangerment and, 27–28; nostalgia and, 48–49
Tiananmen Square massacre (1989), 27, 60, 68
Tourism: Chineseness and, 155; Disney and, 110, 140; ecotourism, 26, 82; Tai O and, 25–26, 37–41, 155; visas, 71–72
Traditional Chinese medicine, 152–53
Traditional ecological knowledge (TEK), 88
Translation: articulation and, 94–96; collaboration and, 90, 93–94, 104; cultural, 95; environmentalists and, 112–16; expert witnesses and, 89–96; as linkage, 176 n. 36; natural, 175 n. 24; place and, 14; political mobilization and, 91; universality and, 14, 95
Tung Chee-hwa, 109, 110, 140, 147
Tung Chung village, 46–47, 51, 137–38

Uniqueness: endangerment and, 27; of Hong Kong, 31–32, 39, 69–70, 71, 148; place-based, 17; of species, 24, 30–31; in Tai O, 26

United Nations Conference on Environment and Development, 88

Universality: cosmopolitanism and, 113; critiques of, 13–14; in Marxism, 144; particularity and, 14, 79–81, 87–90, 95, 104–5, 112, 143, 166–67; poetic thinking and, 168; in political mobilization, 6; of scientific knowledge, 79; skepticism toward, 79–80, 87

Vegetarian Eating Society of Hong Kong, 9

Vocation, 114, 115, 178 n. 8

Waaihgauh, 41. *See also* Nostalgia

Walt Disney Corporation, 140. *See also* Disneyland

Watson, James, 61

Wing-sang Law, 69

Wing Hung. *See* Fan Wing Hung

World Trade Organization (WTO), 49, 62

World Wide Fund for Nature (WWF), 2, 7, 28

World Wildlife Fund. *See* World Wide Fund for Nature

Xi Xi, 139–40

Youngsaye, J. L., 54

Yu Gung Moves the Mountain, 41–42

Yumcha, 61

Yundeichunggin, 44, 45

Yu, Rupert, 79, 81, 90, 92–96, 116–24, 133–34, 156

Zhuangzi, 153

Tim Choy is an associate professor of anthropology
and science and technology studies at the
University of California, Davis.

• • •

Library of Congress Cataloging-in-Publication Data
Choy, Timothy K., 1971–
Ecologies of comparison : an ethnography of
endangerment in Hong Kong / Tim Choy.
p. cm. — (Experimental futures : technological lives,
scientific arts, anthropological voices)
Includes bibliographical references and index.
ISBN 978-0-8223-4931-0 (cloth : alk. paper)
ISBN 978-0-8223-4952-5 (pbk. : alk. paper)
1. Environmentalism—China—Hong Kong.
2. Ethnology—China—Hong Kong. 3. Environmental
protection—Social aspects—China—Hong Kong.
4. Political participation—China—Hong Kong.
I. Title. II. Series: Experimental futures.
GE199.C6C46 2011
333.72095125′09049—dc22   2011010756